AF539202

Vocational Guide to Mobile Communications

Vocational Guide to Mobile Communications

Ankit Tiwari

RANDOM PUBLICATIONS
NEW DELHI (INDIA)

Vocational Guide to Mobile Communications

ISBN 978-93-5111-405-5
© Reserved

All Rights Reserved. No Part of this book may be reproduced in any manner without written permission.

Published in 2014 in India by

RANDOM PUBLICATIONS

4376-A/4B, Gali Murari Lal, Ansari Road
New Delhi-110 002
Phone : +91-11-43580356, +91-11-23289044
e-mail: randomexports@gmail.com, sales@randompublications.com, info@randompublications.com

Type Setting by : Keystoneprintads, Delhi-110051
Digitally Printed at : Replika Press Pvt. Ltd.

Preface

A communication network which does not depend on any physical connection needed to between two communication entities and have the flexibility to be mobile during communication. The current GSM and CDMA technology offers Mobile Communication. Mobile communication allows transmission of voice and multimedia data via a computer or a mobile device without having connected to any physical or fixed link. Mobile communication is evolving day by day and has become a must have for everyone. Mobile communication is the exchange of voice and data using a communication infrastructure at the same time regardless of any physical link.

Mobile Communications and Networking research is focused on next generation cellular wireless data network architectures. New technologies for improving spectral efficiency, spectrum utility and spatial reuse of spectrum on the radio-access network are being developed. On the mobile core network, research is oriented towards efficiently handling the exploding amount of mobile data traffic, enabling new revenue generating services for mobile operators and network CAPEX/OPEX reduction techniques. A variety of approaches including analysis, simulations, modeling, as well as experimental prototype design and network measurements are used to validate the technologies developed.

Mobile Technology has groomed a lot in the past few years, major reasons for rapid advancements in the mobile network technology is requirements for being mobile or connected to moves. Latest mobile handsets offer features which one had never thought off, ultimately it forces mobile network companies to bring these features in practice use to take commercial advantages. Cellular is one of the fastest ongoing growing telecommunication industry in the world. It may easily be one of those industries which may never suffer from the economics of the world.

Wireless communications are a type of data communication that is performed and delivered wirelessly. This is a broad term that incorporates all procedures and forms of connecting and communicating between two or more devices using a wireless signal through wireless communication technologies and devices. Wireless communication is among technology's biggest contributions to mankind. Wireless communication involves the transmission of information over a distance without the help of wires, cables or any other forms of electrical conductors. The transmitted distance can be anywhere between a few meters (for example, a television's remote control) and thousands of kilometres (for example, radio communication).

The book is self-explanatory to a large extent and also serves as convenient reference for engineers working in the Mobile Communications field.

I thank all members of my team who have helped in the preparation of the book. My special thanks go to "Random Publications" who have published the book.

*– **Ankit Tiwari***

Contents

1

Introduction

A communication network (either public or private) which doesn't depend on any physical connection between two communication entities and have flexibility to be mobile during communication. The current GSM and CDMA technologies offers Mobile Communication.

MOBILE COMMUNICATION TECHNOLOGIES

Mobile communication allows transmission of voice and multimedia data via a computer or a mobile device without having connected to any physical or fixed link. Mobile communication is evolving day by day and has become a must have for everyone. Mobile communication is the exchange of voice and data using a communication infrastructure at the same time regardless of any physical link.

Mobile communication technologies not only benefiting businesses to perform their operation faster and efficiently but also raising the standard of human lives. Mobile communication or mobile computing is just the two different names for the ability to use the mobile technology while on the move, most of the portable computers and computing equipment which are particular for the use in stationary place or configuration.

Ever since the population has been increasing, it urges the need for communication. Human beings are constantly busy to build up the easiest and the quickest way to communicate. Advancement in technology as the invention of telephone is one of such examples. Every coming generation is marked with one step ahead. Bigger telephones continuously passes through the modification process till the first mobile phone introduced. This journey of mobile communication technology started from late nineties with the 1st generation mobile technology has now reached till 3rd generation totally changes the canvas of communication mode. Third generation of the mobile phones virtually shrink distant places of the earth into a global village.

There are so many types of mobile computers, such as laptops, PDAs, PDA phones and other mobility devices were introduced in the mid of 1990s including wearable technology as well. And to use these types of mobility equipments we need to use right technology to make it more secure and reliable infrastructure. If we talk about the mobile communication technologies we can count on many mobile technologies available today such as 2G, 3G, 4G, WiMAX, Wibro, EDGE, GPRS and many others.

Mobile computing or mobile communication technologies based on different security standards and as well as the transmission protocol behind it. In this section we have discussed in detail about different mobile computing technologies and mobile communication systems. What makes mobile communication or mobile computing different from other wired and wireless technologies.

2G MOBILE TECHNOLOGY

2G Mobile technology revolutionized the concept of mobile uses as compared to 1G. 1st generation used the mobile with analog system only used for the verbal communication. 2G Technology mobiles are now able to do not only verbal communication but also able to send text.

There are 3 different types of 2G mobile technologies based on the system they are designed. Adoption of any one of the technology makes them somewhat different from the other because of the difference in their working method. All these 2G mobile technologies have different properties to facilitate a special feature of a mobile.

FDMA (2G MOBILE TECHNOLOGY)

It works like a radio system by separating the frequency into equal spectrum but affecting the quality of voice. FDMA is an analog system still exists in 2G mobile technology with the digital module of 2G in limited area. Instead of frequency division now cellular based technology which divide geographical areas not frequency and improve the service. Properties

- This technology is first used in 1G mobile as an analog system, introduced in 2G with the increase in its frequency with the help of cellular technology.
- Able to carry digital transmission but digital transmission is not quality wise as good as in case of analog system.
- Facilitate with the feature of analogue system by enabling the accessibility of call.

TDMA (2G Mobile Technology)

2G Mobiles uses TDMA (Time Division Multiple Access) technology in some of its models. It actually divides the band into three time-periods. TDMA contains technologies GSM (Global Service Mobile Communication), which is the most common technology, uses widely across the world. Properties

- It provides roaming in more than 200 countries. This international roaming feature attracts the subscribers to use it for travelling purpose.

- The most used and the most appreciative feature is that of short messaging service (sms).
- Use of sim is a prominent feature of GSM. It needs only a sim to start communication at a particular region.
- IDEN abbreviated for Integrated Digital Enhanced Network, which is TDMA based technology based on GSM, is utilized by Motorola limited to United State and Canada only.
- D-AMPS (Digital Advanced Mobile Phone System) also known as IS-136 used in America only North and South regions.
- PCS (Personal Communication System) is used in Japan.

CDMA (2G Mobile Technology)

Unlike TDMA, CDMA works using the entire band with the help of code. CDMA is based on a wide spectrum as many calls laid over each other identifying on the basis of unique code.

- CDMA gives a separate code to a separate phone.
- Increase the frequency band space by assigning code in sequence.
- Both senders and receivers are able to use a full band with the help of using their codes.
- Contrast to one analog call, nearly dozen calls can be channelized at the same time.
- IS-95 (Interim Standard) widely uses in Korea increases the capacity as able to use complete band.

2G TECHNOLOGY, SECOND GENERATION TECHNOLOGY

Second generation (2g) telephone technology is based on GSM or in other words global system for mobile communication. Second generation was launched in Finland in the year 1991.

How 2G works, Uses of 2G technology (Second Generation technology)

2G network allows for much greater penetration intensity. 2G technologies enabled the various mobile phone networks to provide the services such as text messages, picture messages and MMS (multi media messages). 2G technology is more efficient. 2G technology holds sufficient security for both the sender and the receiver. All text messages are digitally encrypted. This digital encryption allows for the transfer of data in such a way that only the intended receiver can receive and read it.

2G Technologies (Second Generation Technologies)

Second generation technologies are either time division multiple access

(TDMA) or code division multiple access (CDMA). TDMA allows for the division of signal into time slots. CDMA allocates each user a special code to communicate over a multiplex physical channel. Different TDMA technologies are GSM, PDC, iDEN, iS-136.CDMA technology is IS-95. GSM has its origin from the Group special Mobile, in Europe. GSM (Global system for mobile communication) is the most admired standard of all the mobile technologies. Although this technology originates from the Europe, but now it is used in more than 212 countries in the world. GSM technology was the first one to help establish international roaming. This enabled the mobile subscribers to use their mobile phone connections in many different countries of the world's is based on digital signals,unlike 1G technologies which were used to transfer analogue signals. GSM has enabled the users to make use of the short message services (SMS) to any mobile network at any time. SMS is a cheap and easy way to send a message to anyone, other than the voice call or conference. This technology is beneficial to both the network operators and the ultimate users at the same time.

Another use of this technology is the availability of international emergency numbers, which can be used by international users anytime without having to know the local emergency numbers. PDC or personal digital cellular technology was developed in Japan, and is exclusively used in JAPAN as well. PDC uses 25 KHz frequency. Docomo launched its first digital service of PDC in 1993.integrated digital enhanced network (iDEN) was developed by MOTOROLA, as a major mobile technology. It enabled the mobile users to make use of complex trunked radio and mobile phones. iDEN has a frequency of about 25Khz.i DEN allows three or six user per mobile channel.iS-136 is a second generation cellular phone system. It is also known as digital AMPS. D-AMPS were widely popular in America and Canada. However now it is in the declining phase. This technology is facing a strong competition by GSM technologies. Now the network carriers have adopted GSM and other CDMA 2000 technologies at large. Interim standard 95 is a first and the foremost CDMA cellular technology. It is most famous by its brand name known as cdmaOne. It makes use of the CDMA to transfer the voice signals and data signals from cellular phones to cell sites (cell sites is cellular network).

Benefits of 2G technology (Second Generation)

Digital signals require consume less battery power, so it helps mobile batteries to last long. Digital coding improves the voice clarity and reduces noise in the line. Digital signals are considered environment friendly. The use of digital data service assists mobile network operators to introduce short message service over the cellular phones. Digital encryption has provided secrecy and safety to the data and voice calls. The use of 2G technology requires strong digital signals to help mobile phones work. If there is no network coverage in any specific area, digital signals would be weak.

3G TECHNOLOGY

If you want augmented bandwidth, multiple mobile applications and clarity of digital signals, then3G (Thrid Generation Technology) is your gateway. GSM technology was able to transfer circuit switched data over the network. The use of 3G technology is also able to transmit packet switch data efficiently at better and increased bandwidth. 3G mobile technologies proffers more advanced services to mobile users. It can help many multimedia services to function. The spectral efficiency of 3G technology is better than 2G technologies. Spectral efficiency is the measurement of rate of information transfer over any communication sytem. 3G is also known as IMT-2000.

3G TECHNOLOGY AND 3G TECHNOLOGIES CHARACTERISTICS

3G technologies make use of TDMA and CDMA. 3G (Thrid Generation Technology) technologies make use of value added services like mobile television, GPS (global positioning system) and video conferencing. The basic feature of 3G Technology (Thrid Generation Technology) is fast data transfer rates. However this feature is not currently working properly because, ITU 200 is still making decision to fix the data rates. It is expected that 2mbit/sec for stationary users, while 348kbits when moving or traveling. ITU sell various frequency rates in order to make use of broadband technologies. Network authentication has won the trust of users, because the user can rely on its network as a reliable source of transferring data.3G technology is much flexible, because it is able to support the 5 major radio technologies. These radio technologies operate under CDMA, TDMA and FDMA.CDMA holds for IMT-DS (direct spread), IMT-MC (multi carrier). TDMA accounts for IMT-TC (time code), IMT-SC (single carrier). FDMA has only one radio interface known as IMT-FC or frequency code. Third generation technology is really affordable due to the agreement of industry. This agreement took pace in order to increase its adoption by the users. 3G (Thrid Generation Technology) system is compatible to work with the 2G technologies.3G (Thrid Generation Technology) technologies holds the vision that they should be expandable on demand. The aim of the 3G (Thrid Generation Technology) is to allow for more coverage and growth with minimum investment.

3G technologies (Third Generation Technologies)

There are many 3G technologies as W-CDMA, GSM EDGE, UMTS, DECT, WiMax and CDMA 2000.Enhanced data rates for GSM evolution or EDGE is termed to as a backward digital technology, because it can operate with older devices. EDGE allows for faster data transfer than existing GSM.EDGE was introduced by AT& T in 2003.EDGE has increased the GSM coverage up to three times more. EDGE is a 3G Technology (Thrid Generation Technology); therefore it can be used for packet switched systems. Universal mobile telecommunications systems.UMTS conforms to ITU IMT 2000 standard. It is complex network and allows for covering radio access, core network and USIM (subscriber identity module). It is a relatively expensive technology for the network operators because it requires new and separate infrastructure for its coverage. The GSM is the base of this technology. CDMA is also referred to as IMT-MC. this technology is close to 2G technology GSM because it is also backward compatible. Digital enhanced cordless telecommunications (DECT) is another 3G Technology (Thrid Generation Technology). DETC was developed by European telecommunications standards institute; however it is now widely popular in the other countries of the world as well. It runs over a frequency of 1900Mhz. WiMax is a 3G Technology (Thrid Generation Technology) and it is referred to as worldwide interoperability for microwave access. It is a wireless technology. It transmits variety of wireless signals. It can be operated on the multi point and point modes. it is portable technology. This technology is based on the wireless internet access. Name of the WiMax was developed in 2001 by WiMax forum. This technology removes the need for wires and is capable enough to provide 10mbits/sec. it can connect you to hotspot.

How 3G Technology Works

As every generation has been introduced with a new advanced technology, third generation (3G) is no exception. Latest developments in technology are not stunning now rather people are waiting impatiently for the newest modification and improvements and ready to go for it. 3G technology is the modified form of second generation it is better to say that all the best features of different versions of 2nd generation are combined into third generation. Resultantly 3G Technology is known as Smartphone with high data transfer rate, WiFi hotspots connectivity and multimedia features.

Three important technologies which make the 3G standard are CDMA 2000, WCDMA (UMTS), and TD-SCDMA. The prominent features after the integration of high capability technologies are like high speed data transfer rate at 3Mbps prop up the usage of internet. It's the matter of your need now to which mode you want to switch as to PC, internet, or phone mode, simply it's a 3 in 1. Let see the technologies used in 3G and their function how they work.

How 3G (CDMA 2000) Works

Code Division Multiple Access 2000 is approved by 3GPP2 Organization. CDMA200 hybrid with IS-95 B provides an unlimited access to IMT-200 Band as well as cdma 200 1x and ideal conditions for the highest data transfer rate. The cdma 2000 1x evolves into cdma 200 1x EV. This cdma 200 1x EV IS put into service in two different forms;

- Cdma 2000 1x EV-DO- 1X Evolution data only able to use 1.25 MHz
- Cdma 2000 1x EV-DV- 1x Evolution Data and Voice also use 1.25 MHz

All these versions are supposed to attain the highest speed for greater efficiency of the mobile phones.

How 3G (W-CDMA/ UMTS) Works

3G mobile technology has been marked by the CDMA accomplishment. ETSI Alpha group develop this technology on radio access method. W-CDMA offers challenges in shapes of versatility and complexity of its design. Its multifaceted single algorithm made the complete system more difficult hence the receiver becomes a more complex device. It provides friendly environment to the multi-users with greater simulation and broader interface able to transfer data with time variations.

UMTS network group is experimenting on this new technology to maintain previous 2G module features and the added features as well in 3G.

How 3G (TD-SCDMA) Works

Developed by China Wireless Telecommunication Standard group TD-SCDMA is approved by the ITU. This technology is based on time synchronization with CDMA. This time division is based on duplex approach where uplink and downlink traffic transmit in different time slots. This synchronization gives flexibility to the spectrum for uplink and downlink transmission depending on symmetrical or asymmetrical information. Asymmetrical information comprises e-mail and internet applications while call system comes under symmetrical information. During asymmetrical applications downlink is given preference over uplink. Preference is given to uplink during telephony.

3G Mobile technology is providing 144 kbps connectivity speed which is the highest one in the present era. It is offering wireless broadband facility an entertainment opportunity of downloading music and videos, games with 3D effects and conference calls with video facility.

Based on the services, feature plans and areas with the 3G coverage there are many carriers who are providing it. It is still not fully accessible in all the countries so limitations hinder the availability. The real 3G technology can be enjoyed with a new mobile phone set and the 3G service pack.

All the three technologies mentioned above are working in 3G at their best. Now everyone is looking forward with expected amazement for the 4G standard as they have for sure belief on the further advancement in the field of telecommunication already reaches the peak of cellular technology.

DIFFERENCE BETWEEN 2G AND 3G TECHNOLOGY

Second Generation (2G) technology was launched in the year 1991 in Finland. It is based on the technology known as global system for mobile communication or in short we can say GSM. This technology enabled various networks to provide services like text messages, picture messages and MMS.

In this technology all text messages are digitally encrypted due to which only the intended receiver receives message. These digital signals consume less battery power, so it helps in saving the battery of mobiles.

The technologies used in 2G are either TDMA (Time Division Multiple Access) which divides signal into different time slots or CDMA (Code Division Multiple Access) which allocates a special code to each user so as to communicate over a multiplex physical channel. 3G technology generally refers to the standard of accessibility and speed of mobile devices. It was first used in Japan in the year 2001. The standards of the technology were set by the International Telecommunication Union (ITU). This technology enables use of various services like GPS (Global Positioning System), mobile television and video conferencing. It not only enables them to be used worldwide, but also provides with better bandwidth and increased speed.

This technology is much more flexible as it can support 5 major radio technologies that operate under CDMA, TDMA and FDMA. CDMA accounts for IMT-DS (direct speed), IMT-MC (multi carrier). TDMA holds for IMT-TC (time code), IMT-SC (single carrier). This technology is also comfortable to work with 2G technologies. The main aim of this technology is to allow much better coverage and growth with minimum investment.

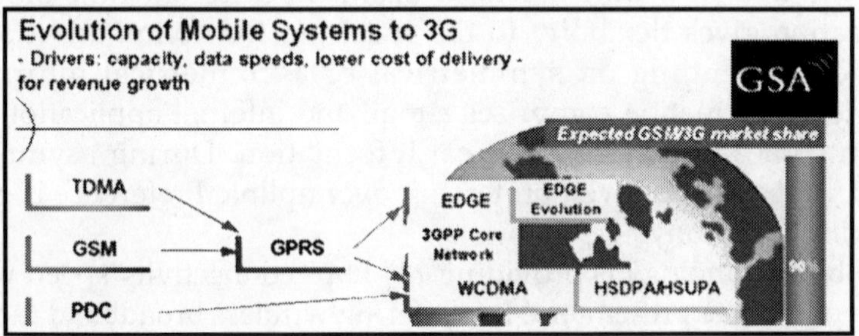

Fig. Evolution of Mobile system from 2G to 3G

Difference between 2G and 3G Technology:

- *Cost:* The license fee to be paid for 3G network is much higher as compared to 2G networks. The network construction and maintenance of 3G is much costlier than 2G networks. Also from the customers point of view the expenditure for 3G network will be excessively high if they make use of the various applications of 3G.
- *Data Transmission:* The main difference between 2G and 3G networks is seen by the mobile users who download data and browse the Interneton the mobile phones. They find much faster download speeds, faster access to the data and applications in 3G networks as compared to 2G networks. 2G networks are less compatible with the functions of smart phone. The speed of data transmission in 2G network is less than 50,000 bits per sec while in 3G it can be more than 4 million bits per sec.

- *Function:* The main function of 2G technology is the transmission of information via voice signals while that of 3G technologies is data transfer via video conferencing, MMS etc.
- *Features:* The features like mobile TV, video transfers and GPS systems are the additional features of 3G technology that are not available with 2G technologies.
- *Frequencies:* 2G technology uses a broad range of frequencies in both upper and lower bands, under which the transmission depends on conditions such as weather. A drawback of 3G is that it is simply not available in certain regions.
- *Implication:* 3G technology offers a high level of security as compared to 2G technology because 3G networks permit validation measures when communicating with other devices.
- *Making Calls:* Calls can be made easily on both 2G and 3G networks with no real noticeable differences except that in 3G network video calls can also be made. The transmission of text messages and photos is available in both the networks but 2G networks have data limit and the speed of the data transmission is also very slow as compared to 3G.
- *Speed:* The downloading and uploading speeds available in 2G technologies are up to 236 Kbps. While in 3G technology the downloading and uploading speeds are up to 21 Mbps and 5.7 Mbps respectively.

4G TECHNOLOGY

WHAT IS 4G TECHNOLOGY

When talking about 4G, question comes to our mind is what is 4G Technology. 4G is short forFourth (4th) Generation Technology. 4G Technology is basically the extension in the 3G technology with more bandwidth and services offers in the 3G. But at this time nobody exactly knows the true 4G definition. Some people say that 4G technology is the future technologies that are mostly in their maturity period. The expectation for the 4G technology is basically the high quality audio/video streaming over end to end Internet Protocol. If the Internet Protocol (IP) multimedia sub-system movement achieves what it going to do, nothing of this possibly will matter. WiMAX or mobile structural design will become progressively more translucent, and therefore the acceptance of several architectures by a particular network operator ever more common.

Many Technologies appear in many different flavours and have many diverse tags attached to them, but that does not really indicate that they are moving in dissimilar tracks. The technologies that fall in the 4G categories are UMTS, OFDM, SDR, TD-SCDMA, MIMO and WiMAX to the some extent.

4G Technology offers high data rates that will generate new trends for the market and prospects for established as well as for new telecommunication businesses. 4G networks, when tied together with mobile phones with in-built higher resolution digital cameras and also High Definition capabilities will facilitate video blogs.

After successful implementation, 4G technology is likely to enable ubiquitous computing, that will simultaneously connects to numerous high date speed networks offers faultless handoffs all over the geographical regions. Many network operators possibly utilize technologies for example; wireless mesh networks and cognitive radio network to guarantee secure connection & competently allocates equally network traffic and bandwidth.

Some of the companies trying 4G mobile communication at 100 Mbps for mobile users and up to 1 Gbps over fixed stations. They planned on publicly launching their first commercial wireless network around 2010. As far as other competitor's mobile communication companies working on 4G technology even more quickly.

Sprint Nextel was planned to launch WiMAX over 4G broadband mobile network in United States. Some of the other developed countries like United Kingdom stated a plan to sale via auction of 4G mobile frequencies couple of years back. The word "MAGIC" also refers to 4G technology which stands for Mobile multimedia, Any-where, Global mobility solutions over, Integrated wireless and Customized services.

With 4G mobile technology around the corner it is difficult to imagine how we were able to go around our everyday lives without mobile communication speeds of up to 100Mbps. Not only will 4G improve mobile communication, it will likely positively impact wireless technology for online gaming and internet browsing in the home. If you want to test the capability of 4G technology in the online gaming industry visit spinpalace.co.uk and download their latest software for the ultimate gaming experience. Visit them on your 4G compatible mobile device and experience the 4G technology.

GSM (GLOBAL SYSTEM FOR MOBILE COMMUNICATION)

- E-Mail
- Print
- A
- AA
- AAA
- inShare
- Facebook
- Twitter
- Share This
- RSS
- Reprints

GSM (Global System for Mobile communication) is a digital mobile telephony system that is widely used in Europe and other parts of the world. GSM uses a variation of time division multiple access (TDMA) and is the most widely used of the three digital wireless telephony technologies (TDMA, GSM, and CDMA). GSM digitizes and compresses data, then sends it down a channel with two other streams of user data, each in its own time slot. It operates at either the 900 MHz or 1800 MHz frequency band.

Mobile services based on GSM technology were first launched in Finland in 1991. Today, more than 690 mobile networks provide GSM services across 213 countries and GSM represents 82.4 per cent of all global mobile connections. According to GSM World, there are now more than 2 billion GSM mobile phone users worldwide. GSM World references China as "the largest single GSM market, with more than 370 million users, followed by Russia with 145 million, India with 83 million and the USA with 78 million users."

Since many GSM network operators have roaming agreements with foreign operators, users can often continue to use their mobile phones when they travel to other countries. SIM cards (Subscriber Identity Module) holding home network access configurations may be switched to those will metered local access, significantly reducing roaming costs while experiencing no reductions in service.

GSM, together with other technologies, is part of the evolution of wireless mobile telemmunications that includes High-Speed Circuit-Switched Data (HSCSD), General Packet Radio System (GPRS), Enhanced Data GSM Environment (EDGE), and Universal Mobile Telecommunications Service (UMTS).

CELLULAR NETWORKS

Every day, we make calls, send messages or connect to the Internet using mobile devices but rarely stop to wonder how it all works. So what is actually going on behind the scenes to enable your mobile to do what it does?

Mobile communications work by using low power radio waves necessary to carry speech and data. When a call is made, the signal is handed across a network of linked geographic areas called cells - hence the term cellphone - until it reaches its destination.

A piece of equipment called a base station transmits signals from one cell to the next, or to land-line networks. Each cell is the area that each base station covers. Base stations are often called masts, towers or cell-sites.

CELLULAR HISTORY

Work began on a European standard for digital cellular telephony in the European Conference of Postal and Telecommunication Administrations (CEPT) in the mid-1980s. The task was entrusted to a committee known as Groupe Spécial Mobile (GSM™), aided by a 'permanent nucleus' of technical support personnel, based in Paris. In 1992, the work was moved to ETSI and the technical committee renamed Special Mobile Group (SMG). The permanent nucleus moved to ETSI headquarters and was renamed Project Team (PT) 12, later Special Task Force (STF) 12 (it was sometimes informally referred to as PT-SMG). In order not to destabilize the development, the document numbering system created by CEPT GSM was retained, though the term 'Recommendation' which CEPT used (this term is still employed by the International Telecommunication Union, ITU) was phased out and replaced by 'Technical Specification' (TS).

GSM specifications

Over the next couple of years, the specifications for the first phase of the new system were completed, and were published by ETSI. These are listed in TS 01.01 (Phase 1). In fact, the 'Technical Specification' did not fit well with the deliverable types produced by ETSI at the time, and the specifications are published as a mixture of GSM Technical Specifications, GTSs (later Technical Specifications, TSs), European Telecommunication Standards, ETSs (later European Standards, ENs) and ETSI Technical Reports, ETRs (later Technical Reports, TRs).

Indeed, for some specifications, it was not unknown for the deliverable type to change when the document was revised. For this reason, when dealing with cellular deliverables, it is easier to refer to the GSM specification number rather than the ETSI deliverable number. However, a software tool is provided to convert from one to the other.

The Phase 1 specifications were used for trials of the GSM system ('GSM' had become a well-known term, but its original connotation was no longer relevant, so it was reborn as an abbreviation for Global System for Mobile communications). Results were fed back into the standardization process to produce a second phase. This phase was widely implemented in commercial networks from the mid-1990s onwards, and work had already begun on a subsequent phase, with further improvements and enhancements.

Data communications

In fact, engineers were already looking beyond the 'second generation' (i.e. GSM) towards a 'third generation', which would employ a different radio access technology to give higher data rates (Earlier analogue technologies were considered to be 'first generation'). GSM had originally been developed purely for voice telephony, but a data capability had been added to cater for a sudden increase in dial-up modem use on fixed lines. A maximum data rate of 9600 bits per second was offered, which matched or indeed exceeded that available over fixed line modems at the time. However, by the mid-1990s, dial-up data rates had surpassed this, and the performance of GSM circuit-switched data connections was now perceived as inadequate. Since a third generation digital cellular system was now envisaged, SMG decided not to call the next phase of the GSM specifications Phase 3: this might have been confused with 3rd Generation. Instead, it was decided to use the term 'Phase 2+', and the specification set was completed in 1996.

GSM Releases

By this time, it was clear that a more or less annual release of specifications, each containing improvements and additional service features over its predecessor, was desirable. New developments could be standardized without destabilizing previous releases (although faults detected with the earlier specifications would continue to be fixed). The nomenclature of the phased releases was now changed: the specification sets were still referred to as 'Phase 2+' but a release year was added. In fact, the releases became known simply as R97, R98, and R99, and the original Phase 2+ became known as R96. The specification sets for all of these Releases are listed in the corresponding Releases of TS 01.01.

Short Message Service

The origin of the text messaging services in GSM lies in the historical development of telecommunication services, and SMS was created by a small group of persons. The work on the standardization of services and the technical realization was approved by the CEPT Groupe Spécial Mobile.

Text messaging was a known telecommunications service years before the development of GSM™ started in 1982. Proposals for text messaging as a service in GSM were made by Nordic, German and French operators, who were all co-operating in the task.

The Nordic operators focussed their work on text messaging by using an access to a message handling system, a service similar to e-mail. This service was standardized by the GSM committee and led to a technical report on the technical realization of the access to Message Handling Systems.

The German and French operators focussed their work on 'Short Message Transmission'. This service uses a dedicated service centre and transmits the text messages over existing signalling paths of the GSM telephony system on a lower priority basis. This transmission method obviously constrains the message to be short: the maximum length, initially estimated as 128 octets, later optimised to 160 characters, is still sufficiently long for most personal or professional purposes.

The first phase of work, from February 1985 to the end of 1986, saw the GSM committee specifying the service features of SMS, with most contributions coming from Germany and France. From 1987 onwards the technical realization of SMS was standardized in a small group called Drafting Group Message Handling. The first Chairman of this group and the editor of the key technical specification were provided by Norway (later replaced by the UK), and technical work was mostly provided by Finland, France, Norway, and the UK. The first phase of the SMS specifications comprised items that included service definition, network architecture, topology and protocols, acknowledgement capabilities, functionality for alerting on messages waiting, time stamping and capabilities of identifying application protocols.

The further evolution of the SMS was standardized in the same small group, led by the UK (where the technical work was mostly provided by the UK). Examples of enhancements from this period are automatic replacement of messages, so-called 'flash SMS' and voicemail icons, followed by colour and picture capabilities, and long SMS.

SMS has become a phenomenal success that has been a surprise to many both inside and outside the communications industry.

All about base stations

Base station sites come in a range of shapes and sizes and are essential to making mobile phones work. They house radio transmitters and receivers that pick up signals sent from your phone and transfer them to your network operator, so that you can be connected.

Base stations receive signals from mobile phones at ground-level or in buildings and send out signals that you receive as calls, messages, data and other mobile services.

Their construction requires:

- Structure for mounting antennas
- Antennas for exchanging signals
- Electronic equipment for processing signals

Antennas are typically placed high above the ground to transmit and receive signals between base stations. Masts and towers are often used to increase height but you might also see antennas at the top of taller buildings. In some situations, less visibly prominent base station designs may be used.

DIFFERENT SITUATIONS, DIFFERENT DESIGNS

There are three main types of base station:

Masts or cell-sites

Tall structures that provide the main network coverage. Free-standing masts may be placed at the road-side, in open spaces or wooded areas. Existing buildings are also used so that new construction work is not always needed. Where possible, the visual and environmental impact is limited by slim-line designs, and by trying to make them blend in with the surroundings by painting them, or carefully placing them on existing structures or among trees.

Microcells

Smaller installations that provide extra capacity and fill in gaps in network coverage. They may be incorporated as building features, or fitted to street-lighting posts or other street features. The visual impact of antennas can be reduced by integrating them with an existing structure.

Other base stations

Small base stations are normally used inside buildings, transport terminals or shopping centers. They use very low power levels, and boost reception where coverage is poor or where there are large numbers of users. Differences in operation mean that 3G networks will need more base stations that are closer to each other, but typically with smaller antennas.

Location, location, location

So, where do mobile communications transmitters get located? Three main factors affect the positioning of base station sites:

- Coverage - the area that a base station can service
- Capacity - the level of user demand for service within an area
- Environment - how a site might affect its surroundings

Each of these is affected by a number of other factors:

Range

Using your phone nearer a base station typically needs a lower signal strength, helping to prolong battery life and reducing radio interference with neighbouring network cells.

Distance

The farther you move away from a transmitter, the weaker the signal. Signals are not more intense directly beneath a transmitter, as some people may believe. Radio waves travel outwards from antennas, rather than upwards and downwards. This is why buildings or trees can obstruct the signals.

Landscape

Environmental features such as hills and trees reduce the strength of radio signals. Towers located on top of hills help boost coverage in such situations.

Population

More base stations are needed in towns and cities, where there are more concentrated numbers of mobile phone users. When base stations are needed closer together, they operate on lower power. All mobile phone signals are controlled and regulated so that they do not interfere with emergency service communications, or TV and radio broadcasting.

2

Mobile Communication and Mobile Networks

Mobile Communications and Networking research is focused on next generation cellular wireless data network architectures. New technologies for improving spectral efficiency, spectrum utility and spatial reuse of spectrum on the radio-access network are being developed. On the mobile core network, research is oriented towards efficiently handling the exploding amount of mobile data traffic, enabling new revenue generating services for mobile operators and network CAPEX/OPEX reduction techniques. Variety of approaches including analysis, simulations, modeling, as well as experimental prototype design and network measurements are used to validate the technologies developed.

Since the implementation of economic reforms and the policy of opening up to the outside world from early 1980s onward, the internal migrations from the western and central regions of the country to the eastern coastal region have been ceaseless. The floating population, predominantly migrant workers, has increased sharply from 30 million in the early 1980s to 140 million in the early 2000s. Migrant workers in Guangdong are largely from rural areas in the central or western regions of China, such as Hunan, Sichuan, and Guangxi, where a patrilineal structure is still prevalent in the villages.

Some observations suggest that the drastic economic reforms in the countryside have had a substantial impact on the patrilineal structure. Young villagers, as wage-laborers, are economically independent and some even make a significant contribution to the household's income, giving rise to a new balance of power in the family, where the autonomy of the individual has increased considerably. The out-and-return migration processes have also constituted the change of values and goals of migrants and potential migrants. According to our study, the development of new communication technologies such as the growing popularity of television sets in the villages, the adoption of landline phones, and the recent rapid penetration of mobile phones have made the villagers' homes in the less-developed regions more open to developments in the cities of the coastal regions. In other words, the new communication technologies have been conducive to bringing closer the coastal with the central and western regions. New forms of city lives and new values and ideologies have penetrated insensibly through these new technologies to the less-developed villages. This would undoubtedly have shaken the deep-rooted tradition of patriarchy, and have had an effect on the young and on potential migrants.

Indeed, large numbers of young rural villagers are flocking to Guangdong, particularly to the Pearl River Delta, not only to hunt for jobs to improve their standard of living, but also to experience the city life. As migrant workers, these young villagers believe that when they leave their home villages they will also be free from traditional cultural and social fetters. One informant told us precisely in the interview that:

When we first came here (Guangdong), we felt that we were freer than before. In the past (in the home village), the standards of measurement, the judgments, they were all the same.

When they leave their home villages to seek their fortunes in Guangdong, these young migrants do not find life to be easy. Their labor is exploited, they often encounter hostility from local workers or villagers, and the feelings of desolation and insecurity they experience by being in a strange place have caused some to return home. But for those who are determined to stay in Guangdong, contacting their families in their places of origin and joining their kinship network in Guangdong provides important emotional and social support during their stay. Mobile phones, of course, have becomes a necessary tool in maintaining these existing networks in expanded spatiotem-poral contexts.

In the past, without mobile phones, maintaining connections with families in their home villages through fixed phones was very inconvenient. Some migrant workers said that not all the rooms in the dormitories had fixed phones. Even when landline phones were installed, they were always engaged. During important festivals, when the workers call home to extend their seasonal greetings, huge crowds always form around the public phones in the factories and on the streets. Having mobile phones allows migrant workers to connect with their families more conveniently. A male worker said he had felt safer emotionally since he bought his mobile phone because he could maintain regular contact with his family in his home village. Also, a female worker said:

When I want to talk to my mother (at home village), I would use my mobile phone to call her. It is not that expensive—just 0.25 yuan I always call back home. In fact, there isn't anything important. I just want to chat (with them).

Mobile phones can provide them the immediate absent presence of their families far away in their home village when workers are feeling lonely and thinking of them.

In the past, it was also very inconvenient for workers to maintain contact with their relatives even when they were working in factories nearby. When they wanted to meet their relatives, what a worker could do was go in person to the factories where their relatives worked and wait outside before they got off work. A mobile phone can reduce these difficulties and allow workers to contact their relatives to arrange gatherings freely and easily. If their kinsmen are scattered far and wide in Guangdong and meetings can barely be held regularly, they can connect either by making a call or by sending an SMS. There is an interesting story about a young migrant worker from Guangxi that demonstrates the strength of the connectivity of mobile phones. The worker had a job arranged for her by her father in their home village but she slipped and worked in Guangzhou City alone. Eventually she was caught because her father tracked her mobile phone and she was located by relatives. She escaped again and has now been working in Dongguan. She told us:

I don't want to have mobile phone anymore otherwise everyone can find you out easily. I just don't want to be found by my father.

Although a mobile phone is a personalized mobile device for the migrant workers, it is indeed a very powerful tool in building mobile cyber-kinship networks in Guangdong. If a large portion of the inhabitants of a village are working in a coastal region, which is a common phenomenon, and if they have become connected via mobile phones, a strong mobile cyber-kinship network would be formed there.

Mobile phones are not only helping migrants to maintain existing kinship relationships in expanded spatiotemporal contexts, but also to prolong new social relationships developed in the workplace. Workers come to know each other when working on the same assembly lines or staying in the same dormitories. They develop friendships among workers from different villages, towns, or provinces even though they are not kinsmen and would normally consider each other to be "outsiders". In traditional kinship relationships kinsmen are denned as "insiders" upon whom one can lay one's trust; others are "outsiders" and have to be measured in instrumental terms. Thus, industrialization in Guangdong provides a platform for developing a new kind of social relationship where people are not simply polarized into either "insiders" or "outsiders." Migrant workers are nevertheless highly mobile and most of them do not stay in one factory for a long period of time. Once they find other factories offering higher incomes and better welfare provisions, they will leave for these factories immediately. Thus in the past, friendships developed would rarely be prolonged when workers left the factories. This situation recently has been changed substantially. A worker told us that he could maintain regular contact with his good friends through mobile phone:

We send SMS to friends for maintaining contact. When we receive messages that means our friends are still thinking of us. Three days (without contact) would be okay. If I don't receive anything from them for a while, and I don't know what they are doing, I will send SMS to them. We have to keep the connection, and if we lose contact, it is just like a thread broken.

Another worker said, "When you received SMS from your friend, just one or two words, you would feel happy." He further maintained that he could develop his own circle of friends after he had gotten a mobile phone.

Mobile phones, which free individuals from temporal and physical constraints, serve an important function of preserving this new kind of social network among the migrant workers with different places of origins.

With expanded social relationships and new ways of connecting, migrant workers are more resourceful in obtaining information. By simply making calls to their kinsmen and friends, migrant workers can more easily learn about job opportunities in Guangdong. These networks are essential in empowering the workers when they are facing difficulties. The following section discusses how these networks could empower the workers and subsequently increase the rate of mobility of the migrant workers, both directly and indirectly.

In Dongguan, formal channels for accessing information about the job market were very limited in the past and finding a job was very inconvenient. Each factory would post vacancies on their front doors when they wanted to recruit workers, so young villagers had to go around to the factories in each village to see whether there were vacancies. If these villagers had relatives working in Guangdong, it was relatively easier for them to gain access to information about the job market; but even if their relatives found a job vacancy for them they sometimes missed their chance when they were traveling around Guangdong and could not be reached immediately if they did not have pagers or mobiles. Now they have more convenient access to job information than ever before when they have mobile phones. A worker described what he usually did when his kinsmen called him:

Usually they send me an SMS from the home village and inquire whether there are chances, and if the answer is positive they will make a voice call to find out more about the details. If they have already come to Dongguan, then I will call my friends or send SMSs and see whether there are any factories that need workers. If there are, then I will accompany them to the factory or give them my friend's mobile phone number for them to make the connection.

In the past, when migrant workers had difficulty gaining access to information about the job market, they would not readily give up their jobs because finding another one, and particularly a better one, was not an easy task.

If they were fired they had to go back to their home village. Thus, when faced with poor treatment they had to be humble and submissive, or even to make abject apologies to the factory management. Otherwise, they would incur more unfair treatment. If they logged complaints against maltreatment, they would be violently assaulted by the factory guards and eventually dismissed. The assaulting of migrant workers has been a common phenomenon.

Now, mobile phones have strengthened the bargaining power of migrant workers with their factory proprietors. For example, there is a story concerning a fight between two workers. One was injured and demanded that the other compensate him for his injuries. After investigating the fight, the factory manager supported the injured party and asked the other to pay 200 yuan toward medical costs or he would be fired. The worker refused to pay the compensation and left the factory without any hesitation. Before he left, he reminded the manager to call his mobile phone to tell him if he was due any wages so that he could come back and collect them. The factory manager told us that the worker was still staying in the village because he had found a job in another factory soon after being fired. In this case, the manager felt that it was more difficult to control the workers, as they are no longer afraid of being fired. He pointed out that it is precisely the penetration of mobile phones among the workers that has imperceptibly empowered them to a great extent.

Another factory proprietor, whose business is producing garments, told us that he was very afraid of the use of mobile phones among the workers in his factory. He said that, for instance, during the lunch break workers can use SMSs to share information about the salaries, benefits, promotion opportunities, and working conditions of other factories. Once they discover that any of these conditions are better at another factory, they will quit their jobs immediately. They will introduce their relatives and fellow villagers to this factory as well. Usually, there will be a chain effect. One worker resigns and goes to the other factory, and his or her relatives and fellow villagers will also resign and join that worker in that factory. The proprietor said that one time he had more than seven workers, all belonging to one family, leave his factory after receiving SMSs during the lunch break. He maintained that he really had increased the basic salary of his workers; for instance, the salary of the skilled workers already had risen to 1,000 yuan, which is a very good income for garment workers in Dongguan. Even though the workers were happy with the salary, they demanded more holidays and less overtime work. To satisfy as many of their demands as possible, his factory now needs more workers than before. He complained:

If they have less information about the job market, they will be less likely to move around Even if we install some kind of interference technology they can still call their kinsmen or friends after work. The only way to minimize their contacts with others is to move our factory to a place where mobile phones cannot receive any signals at all. This proprietor believed the adoption of mobile phones among the workers has been a key factor affecting mobility rates in Dongguan, even though he knows that the shortage of labor in Guangdong has made it easier than before for workers to be mobile, as he is very close to the workers and understands their daily life very well. As networks have expanded and connectivity has become highly efficient, the flow of job market information has also become rapid and extensive. This, in turn, has provided abundant information about job markets that is extremely easy for workers to access nowadays. Mobile phones have helped workers fight for their rights more successfully than they were able to in the past.

MOBILE PHONE NETWORKS

BASE STATIONS AND HANDSETS

A mobile phone sends and receives information (voice messages, fax, computer data, etc) by radio communication. Radio frequency signals are transmitted from the phone to the nearest base station and incoming signals (carrying the speech from the person to whom the phone user is listening) are sent from the base station to the phone at a slightly different frequency. Base stations link mobile phones to the rest of the mobile and fixed phone network. Once the signal reaches a base station it can be transmitted to the main telephone network, either by telephone cables or by higher frequency radio links between an antenna (e.g. dish) at the base station and another at a terminal connected to the main telephone network.

'Cellular' Radio

Each base station provides radio coverage to a geographical area known as a cell. Base stations are connected to one another by central switching centres, which track calls and transfer them as the caller moves from one cell to the next. The cell structure of a mobile phone network[1]. An ideal network may be envisaged as consisting of a mesh of hexagonal cells, each with a base station at its centre. The cells overlap at the edges to ensure the mobile phone users always remain within range of a base station. Without sufficient base stations in the right locations, mobile phones will not work.

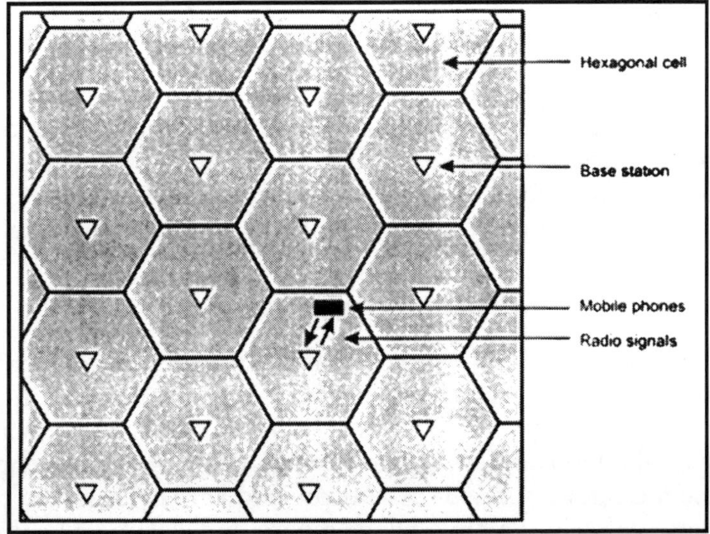

Fig. 'Cellular' Radio

The size of each cell depends on three factors. First, the local terrain; radio signals are blocked by trees, hills and buildings. Second, the frequency band in which the network operates (in general, the higher the radio frequency, the smaller the cell). Third, the capacity (i.e. number of calls) needed in any given area. Base stations are typically spaced about 0.2-0.5 km in towns and 2-5 km apart in the countryside.

If a person with a mobile phone starts to moves out of one cell and into another, the controlling network hands over communications to the adjacent base station.

Why are so many base stations required?

Transmitted signal strength falls off rapidly with distance from base stations, and mobile phones require a certain minimum signal strength to ensure adequate reception. The current generation of GSM base stations cannot communicate over distances greater than 35 km because the delay in receiving radio signals becomes too great. However, the decline of signal strength with distance places a practical limit on coverage of around 10 km. For these reasons an extensive network of base stations is needed to ensure coverage throughout the UK.

Why can't one base station serve my town?

Radio spectrum is a precious natural resource with many different demands upon it (for example, radio and TV broadcasting, emergency communication, navigation aids etc). Consequently the amount made available to each mobile phone operator is limited and this means base stations can only carry a limited number of calls at any one time.

To accommodate the steadily increasing volume of users, network operators have to use the limited number of radio frequencies licensed to them to support the maximum number of mobile phone users. This is achieved by re-using any given radio frequency many times in a network and carefully controlling base station power so that signals arising in different parts of the network do not interfere with each other. The cells are grouped into clusters, with the frequencies allocated to a particular cell within a cluster not being re-used until the corresponding cell in adjacent clusters. This gives a repeating pattern of cells and clusters which can be expanded to provide national coverage.

MOBILE PHONE MANIA

Despite being an urban poor community, Sitio San Roque managed to keep pace with the times—setting up the necessary infrastructure and support services for the mobile phone users among its residents. More than 50 per cent of the local establishments sold not only foodstuff and household items but also mobile phone accessories and prepaid cards. They also provided technical services such as mobile phone repairs and electronic loading or "e-loading". The informants' initiation to the mobile phone resulted from either personal or mediated communication, or from both. They became aware of the mobile phone through the media, particularly the TV. They often saw their favorite soap opera characters, actors, and actresses holding a mobile phone. They also saw numerous advertisements of different brands of mobile phones during the commercial breaks in TV programs.

Almost half the informants saw the mobile phone gadgets in the hands of friends or acquaintances who were members of rich families. They also witnessed rich persons in malls brandishing mobile phones. As a result, some of the informants found themselves yearning for the same gadgets.

Significantly, the informants interacted with people from the upper economic and social classes. Urban poor in-school youth intermingled with wealthier peers. The women entrepreneurs visited affluent families to sell their wares and services, such as laundry, sewing, manicure/pedicure, and house cleaning. Men who worked in construction or repairs almost invariably had rich people among their customers. The rich, who bought expensive gadgets, were described by one focus group discussion (FGD) participant as "angry with their money. They simply don't know what to do with, or how to get rid of, their money".

THE URBAN POOR'S ACQUISITION OF MOBILE PHONES

The high rate of mobile phone acquisition in Sitio San Roque that I found demonstrates that one's economic situation does not necessarily pre-empt one's ability to be technologically equipped. All the FGD participants owned and used mobile phones. Apart from the mobile phone, they also managed to acquire a range of popular technologies such as radios, TVs, karaoke machines, video compact discs (VCDs), DVDs, computers, and the Internet. It had taken them some time to acquire the hardware, but the informants managed somehow to keep abreast with their rich counterparts.

Urban poor informants got their mobile phones through either installment or cash payments. Regardless of the mobile phone's condition (new, used, upgraded, or reconditioned), by far the most common path was by installment payments. Sometimes units were purchased from an enterprising friend, who, for a 20 per cent profit margin, would be paid in four equal installments. The installment system was known locally as *paiyakan* ("shedding of tears" due to the difficulty in collecting payments).

The installment method seems suited to the coping mechanisms of the urban poor, who rely mainly on daily wages for their subsistence. While costly, this mode enables them to satisfy a real or imagined need. Not surprisingly, therefore, the urban poor accumulate debts from all their possible sources of credit.

Cash was usually used to purchase second-hand units. Buyers found these relatively cheap (at USD 28 to 38 per unit) but failed to consider, or simply ignored, the aspects of the phone's quality or longevity. The few who managed to buy brand new mobile phones with cash actually did so because of financial support from overseas foreign workers (OFW), usually parents or rich relatives. Otherwise, they scrimped on basic needs to save money for a mobile phone instead of saving the money or using it for basic household needs.

One avenue by which mobile phone ownership is pursued is by lotteries or raffles. Feeling sure that they would never acquire a mobile phone if they relied merely on their meager wages, some informants tried their luck with lottery. One of them won a new mobile phone at a Christmas party raffle. Although it was an isolated incident, its impact on the raffle winner and his neighbors, who merely learned about it, lingered on. For several months, the superstitious among them continued to bet in lotteries or buy raffle tickets, hoping that Lady Luck would also favor them.

Finally, pawn and trade are used in the acquisition process. This practice suggests the economic usefulness of the mobile phone when used as collateral for a sum of money in times of emergency. Failure to repay it gives the pawnee the right to own the mobile phone. Some of the informants regarded the mobile phone as an investment, not unlike jewelry, that could be pawned for cash. A youth informant would pawn his mobile phone whenever he needed emergency funds for school enrollment fees: "Of course, since we live a hand-to-mouth existence, I wouldn't have money for matriculation. But the cell phone has a value because I can pawn it. There was an instance when I couldn't redeem it due to lack of funds, so I simply sold it".

Engendering the Mobile Phones

Gendered differences, women and men informants said, appeared in the way they actually used the mobile phone:

- Women were talkers and big spenders, while men used the mobile phone for important things and provided the funds for the phone and cell card purchases. Some male informants claimed that women's nature to gossip resulted in big expenses on prepaid cards. They said, "Women are fond of visiting and talking with neighbors, since there is not much to do at home. They always send text messages. That is why they spend a lot on cell cards. Anyway, it is better that they simply stay at home rather than go to the neighbor's house".

  In addition, the men credited the relatively big expenditure on mobile phones to the accessories and the constant replacement of screen logos, ringtones, backlights, cases or pouches, colorful straps, ornamental holders, and even SIM cards. Unlike the women, the men did not have adornments on their mobile phones and simply put them in their pockets.

  Another reason cited by the men, in jest, was that women "have two mouths," that is, that the women were fond of idle talk. The women, however, regarded the frequent use of the mobile phone as socializing and news updating as opposed to rumormongering. One of them countered:

The thing is, whenever we talk with someone, they would immediately conclude that we were engaged in gossip-mongering. For us women, that is not gossiping but rather small talk—a form of enhancing friendship. We're only exchanging or sharing news about what's happening in our place, or with our relatives or other friends in the province or abroad. We're also greeting one another whenever one is celebrating his/her birthday, or when it's Christmas time, the New Year, or Valentine's Day.

One of the FGD participants discovered an important use of the cell phone— counseling someone who was emotionally unstable. Through the cell phone, she said, she was able to provide comfort and advice to one such person. The informant would regularly send her inspirational and religious messages. She claimed that the person was now closer spiritually to the Lord. She proudly related, "I would keep sending inspirational messages to her about the Lord. We used the cell phone so that I could advise her. Otherwise, something might already have happened to her".

The male informants attributed the women's overly solicitous text messages to lack of trust. The women informants did not deny this allegation because some of them had had sad experiences of their husbands' infidelity.

- The women informants were mainly responsible for their children, while the men, as authority figures at home, remained aloof. Some male informants believed that women were "created" to be nurturers of their children and husbands. Moreover, the men said that it was fitting and proper for the women, being the mothers, to take care of the children as well as contact them through text messaging. This was one reason why, at times, they could own a mobile phone ahead of their husbands.
- Women were home mainstays, while the men were engrossed in community concerns and their job searches. Because of this belief, women needed the mobile phone to have someone to talk and consult with. The belief cut the women from the outside world and implied the lack of exposure to events and world-views. This evokes the view that women's own limited experiences imprison their consciousness and lessen their world-making activity. Information media, particularly the mobile phone, can provide a solution to this problem. The mobile phone's capacity to allow women to form their own world-views enables them to expand their network and sustain social or, at times, even romantic relationships.

Reminiscent of Moores's study of TV, the women use the mobile phone as their "lifeline contact with the world". Like the TV, the mobile phone in its turn has become "... an integral part of their day" and combats the loneliness felt by the housewives as a result of their confinement to domestic space.

On the other hand, the male informants contended that as breadwinners they had better things to do. Having no steady jobs, they were engrossed in perpetual job hunting, and were away from home most of the time, thus leaving the women to take care of the house and the children.

- Women had peripheral concerns while the men dealt with essential issues such as community and family welfare. Some of the FGD discussants claimed that the mobile phone should be used only for important matters, referring to their own roles and functions as decision-makers and breadwinners, for example, when they discussed issues with important personalities. They even criticized the women for using the mobile phone for unimportant or frivolous matters, referring to their idle talk and their husband-monitoring activities. The men informants claimed: "Women's use of the cell phone or messages sent is not that important—these are used mostly for idle talk or to monitor the whereabouts of their husbands. This is especially true for the jealous type among them".

  The above shows the men's lack of appreciation for the wives' sociability and for the concern they show for their husbands. It also indicates that the husbands regard the wives' sociability and concern primarily as engaging in *tsismis* (gossips), and as distrusting their husbands.
- The women informants seemed insecure and the men were constrained to reassure them of their love through text messaging. The women manifested this insecurity by sending text messages to their husbands more frequently than the husbands did to them and their children. Thus, the husbands would send texts messages to their wives to persuade them against feeling neglected and jealous.

Unless necessary, the husbands would hardly send text messages to their children, but they regularly sent text messages to their wives because most of them felt obligated to do so, especially when their wives had sent a message telling them to *Tx bak,* Text a reply, or *Sagot ka,* Answer this. Jokingly, some of them would interpret the message as *Sagot ka* to mean *Lagot ka,* You'd better answer this, or else... A husband informant divulged, "But with my wife, there's no way that I won't text her. I'm constrained to do so. Otherwise, she might think that I'm disregarding her and she might get jealous".

Improving communication or relationships with their wives was not the husbands' primary concern for acquiring the mobile phone. But husbands did claim that using the mobile phone for emergency or business purposes actually demonstrated genuine concern for their wives and children. One of them said, "Of course, I love my wife and child, and they are the reasons why I go to work. And that's why I bought a cell phone—so that I can go home quickly whenever I'm needed there. It's important to maintain in contact with them for their own protection".

There also seemed to be a sense of insecurity on the part of the women informants. Perhaps, this insecurity—or feeling left out and lonely—is because they were not formally employed and simply spent most of their time at home. When boredom crept in, the wives would turn to the mobile phone to communicate with their husbands, children, relatives, and friends. This partly explains why the mobile phone has become very important to them.

Women said the reason for regularly sending text messages to their husbands was due to jealousy or the suspicion that their husbands were fooling around with other women. Some of the informants nonchalantly admitted: "Surely, this couldn't be avoided. After all, these are men. Sometimes, the women would seduce them. Well, you know, they are not gays or homosexuals".

Indicating a persistent double-standard morality, the women expressed tolerance and even rationalized the right of the men to be unfaithful. The wives said that they—"because they are women"—should stick to only one partner, thus tacitly underscoring their notion that men are polygamous by nature and thus enhanced their husbands' chauvinism.

SECURITY IN MOBILE NETWORKS

FAULT-TOLERANCE PRINCIPLES FOR SECURITY

We have worked on the development of concepts such as consistency, atomicity, durability, availability, rollback, check points, adaptability etc..

Recently there has been much focus on building secure distributed systems. We note that many of the ideas, concepts, algorithms being proposed in security have many common threads with reliability. To increase reliability in distributed systems, the use of quorums allows the transactions to read and write replica even if some replicas have failed or are unavailable. The systems manage the replicas so that a forum can be formed in the presence of failures. To make systems secure against unauthorized access, one can use the reverse strategy of making it difficult to form quorums. All accesses require permission from a group of authorities who could coordinate to deny a *yes* majority vote.

Checkpointing research has similarities to the work in intrusion-detection. In both cases, either failures or security violations are recorded. The checkpoints ensure that the systems can be brought to a safe consistent state through the use of recovery lines. Such checkpoints can be used to determine secure and safe states of a system. The action taken to rollback to a consistent state will be similar to bring the system to a secure status.

To deal with failures, we build systems that are adaptable. This way, we can deal with the type, duration, severity, timing, extent of a failure. The system will dynamically reconfigure and utilize the best scheme to deal with a specific situation. We must build systems adaptable to security attacks in the same way. The models, experiments, and infrastructure of adaptability to failures are very similar to the ones needed for adaptable secure systems.

There is no way that we can make a system one hundred per cent reliable or secure. In the past, we have designed schemes that deal with one failure and integrated such schemes to build reliable systems. We actually believe that failures will come and go just like a person can get sick and healthy. We can not worry about each individual failures and spend all our resources in dealing with it. We need to identify transient and non catastrophic errors and failures and ignore them if it can benefit the system in dealing with severe causes of non availability. In the same tune, we need to conduct research in dealing with benign security violations that are part of daily system activity. In addition, we must find optimal solutions that allow the applications to succeed inspite of a large mix of failures, security attacks when large number of processes are communicating and accessing large databases.

Such effort is expected to lead us towards a dependable computing system that is adaptable to meet the performance, reliability and security requirements.

Fault-tolerant authentication

The military is greatly expanding their use of wireless networking for the battlefield of twenty-first century based on commercial technology including the IETF Mobile IP protocol suite. Mobile IP allows mobile hosts and mobile routers to change their point of attachment to the network while maintaining continuous network connectivity. Mobile IP does not provide enough security support for tactical military mobile networks. In a tactical military enviornment, the moving components include mobile hosts, mobile base stations (mobile routers/agents), mobile subnets, and even an entire intranet. In a battlefield, when troops move from one place to another, communication networks move with them. The troops need to quickly establish their wireless mobile communication networks in the new location. Most existing wireless network models assume the presence of stationary base stations, which are connected by a high bandwidth wireline network backbone. Wireless network architectures with fixed base stations are unable to adapt to battlefield's dynamic nature.

In a battlefield, fixed base stations are attractive targets, therefore, highly vulnerable. A destruction of a base station will disrupt many communication sessions. Hence, there is a need for mobile computing systems with mobile base stations. For tactical military networks, we can mount base stations on mobile platforms like helicopters and tanks. As the troops move these platforms, they can move their networks with them and provide continuous services to their infantry (mobile hosts).

The tactical military network architecture is a hierarchical arrangement of mobile components. Each unit (corps, division, brigade, battalion and company) has a local area network (or an intranet). High level network interconnections may be based on wired, such as ATM, or satellite. Low level network interconnections are based on wireless links, such as Radio Access Point (RAP) networks. RAPs are highly mobile and support multimedia hosts with mobile TCP/IP. Mobile network introduces several unique network security problems. We discuss the following problems:

Deficiency in mobile IP authentication approaches

In Mobile IP networks, one of the primary security concern is authentication. More specifically, we must implement some mechanisms to allow mobile host and the base station (mobile agent) to authenticate each other as the mobile host moves from cell to cell. Authentication protects the base stations from unauthorized intrusion. One serious deficiency in Mobile IP authentication approaches is that a mobile host (MH) authentication is only through its home agent (HA). If an HA is out of service because of failure, destruction, or temporary non-availability, then all its home mobile hosts will become homeless and will not be able to connect to any other mobile node. Research in fault-tolerant authentication is detailed in.

Deficiency in mobile IP key management

Once a mobile host has been authenticated, it can communicate to other mobile hosts. To secure the communications, data packets should be encrypted before sending and decrypted after receiving. Data privacy protects data transmitted over a communication channel from being either faked or snooped by an unauthorized entity. It, therefore, prevents both active and passive intrusions. Most methods available to enforce authentication, data privacy, integrity and nonrepudiation use some form of cryptography, which requires exchange of secret keys and/or public keys between message sender and receiver. Session key establishment, agility, distribution and management are challenging tasks. Mobile IP, however, does not provide a mechanism for peer-to-peer session key management or for multicast session key management, except assuming manual key distributions. Each key must be kept and distributed in a secure manner. A truly viable key distribution algorithm for mobile network must scale well to a large number of nodes and must be secure and dynamic.

Replicated data consistency

In a battlefield, mobile nodes may be lost, destroyed, or out of order. To provide fault tolerance capability, important data is replicated in many nodes. However, if multiple replicated databases are modified independently, these replicas may not be mutually consistent. Moreover, if different replicas of a data unit are in mutually inconsistent states, the consequences of the inconsistency may depend on the nature of data. For example, in some case weak *reads* may be permitted to retrieve out-of-date information or to retrieve values written by *write* operations that have yet to be committed. To keep data privacy in wireless networks, we can use the same algorithms used in wireline networks. In the following, we identify several research questions and experiments and evaluationd.

Experiments and Evaluation

We need to conduct experiments that simulate the initial key exchange and also the key maintenance between mobile base stations and the mobile hosts. Our experiments simulate the initial key exchange and the key maintenance between mobile base stations and the mobile hosts. Techniques like ISAKMP/Oakley and Diffie Helman will be used for this purpose. We evaluate factors that affect secret key sharing. We use the Berkeley's network simulation tool, ns-2 (network simulator). We briefly outline the following set of experiments for secure mobile systems.

- The simulation environment used for this research achieves fault-tolerant authentication using a hierarchical organization of agents granting service to mobile hosts that are present in its leaf nodes. Five factors come into play when computing the priority of a secret key shared between a leaf node and an internal node. These factors are communication delays between nodes, the processing speed of the internal node, how many times the secret key has been used, the lifetime of that key, and the availability of the key to that internal node.

  We studied the experiments for each factor in. One popular form of analysis is maximizing the effectiveness of the factor being tested, while minimizing the other factors. The tests center around the total time it takes for the system to process a number of requests. The effectiveness of the factor being tested is maximized, and all other effects can be said to be statistically negligible. Each factor is individually tested using different sizes of trees (from 2 levels to 10 levels) to show that any trend holds; there is no correlation between number of levels and service time in this section, since the simulation is reconstructed at each point of experimentation.
- The Intrusion Detection System available in the CERIAS Laboratory can be used to simulate an internal base station setup and to study methods of detecting and acting against possible intrusions into the mobile base stations.

- Additional experimental studies include experiments on hierarchical mobile host authentication, key management in group communications, security as a QoS parameter, and denial of service.

GLOBAL SYSTEM FOR MOBILE COMMUNICATION (GSM)

GSM's air interface is based on narrowband TDMA technology, where available frequency bands are divided into time slots, with each user having access to one time slot at regular intervals. Narrow band TDMA allows eight simultaneous communications on a single 200Khz carrier and is designed to support 16 half-rate channels.

The fundamental unit of time in this TDMA scheme is called a *burst period* and it lasts 15/26 ms (or approx. 0.577 ms). Eight burst periods are grouped into a *TDMA frame* (120/26 ms, or approx. 4.615 ms), which forms the basic unit for the definition of logical channels. One physical channel is one burst period per TDMA frame. A GSM mobile can seamlessly roam nationally and internationally, which requires that registration, authentication, call routing and location updating functions exist and be standardized in GSM networks.

GSM offers a variety of data services. GSM users can send and receive data, at rates up to 9600 bps, to users on POTS (Plain Old Telephone Service), ISDN, Packet Switched Public Data Networks, and Circuit Switched Public Data Networks using a variety of access methods and protocols, such as X.25 or X.32. Other data services include Group 3 facsimile, as described in ITU-T recommendation T.30, which is supported by use of an appropriate fax adapter.

A unique feature of GSM, not found in older analog systems, is the Short Message Service (SMS). SMS is a bi-directional service for short alphanumeric (up to 160 bytes) messages. Messages are transported in a store-and-forward fashion. For point-to-point SMS, a message can be sent to another subscriber to the service, and an acknowledgment of receipt is provided to the sender. SMS can also be used in a cell-broadcast mode, for sending messages such as traffic updates or news updates. Messages can also be stored in the SIM card for later retrieval.

The European version of GSM operates at the 900 MHz frequency (and now at the newer 1800 MHz frequency). Since the North American version of GSM operates at the 1900 MHz frequency, the phones are not interoperable, but the SIMs are. Dual-band 900 -1800 and 900 -1900 phones are already released and in production. Tri-band 900 -1800 -1900 GSM phone are expected to be manufactured in the next few years, which will allow interoperability between Europe and North America A GSM network consists of mobile stations talking to the base transceiver station, on the Um interface. Many BTS are connected to a BSC via the Abis interface and the BSC connect to the MSC (The core switching network) via the A interface.

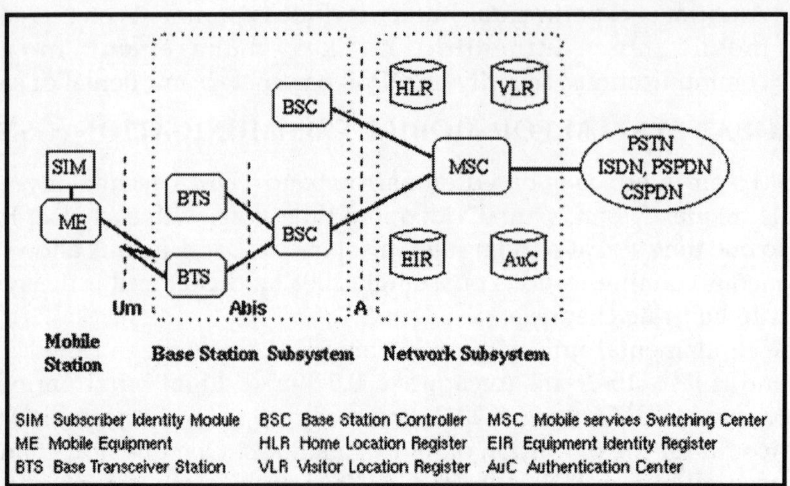

HLR and VLR provide customized subscriber services and allow seamless movement from one cell to another. The Authentication register and the equipment register provide security and authentication. An OMC and a cell broadcast center allow configuration of the network and provide the cell broadcast service in the GSM network (not shown in the diagram).The voice transmitted on the air interface can be encrypted. The speech is coded at 13kbps over the air interface. Using EFR (Enhanced Fullrate Coding) the voice quality approaches the land line quality. Recent developments like AMR (adaptive multi-rate coding) allow speech coding and channel coding to be dynamically adjusted giving acceptable performance even in case of bad radio conditions. The GSM network supports automatic handovers. Since the mobiles are not transmitting or receiving at all times battery consumption can be conserved. Further using DTX and DRX (Discontinuous transmission and reception, mobile transmits or receives only when there is a voice activity detection) batter power can be conserved even more - a highly desirable characteristic of any mobile system. Also since the mobile is not transmitting or receiving at all times, this allows the mobile to listen to control channels and to provide useful information about other channels back to the cell.

Recent developments and initiatives include:

- GSM Association together with the Universal Wireless Communications Consortium (UWCC), which represents the interests of the TDMA community, are working towards inter-standard roaming between GSM and TDMA (ANSI-136) networks.
- The majority of European GSM operators plan to implement general packet radio system (GPRS) technology as their network evolution path to third-generation
- MExE will allow operators to provide customized, user-friendly interfaces to a host of services from GSM, through GPRS and eventually UMTS. The first implementations of MExE are expected

to support the wireless application protocol (WAP) and Java applications. MExE can extend the capabilities that currently exist within WAP by enabling a more flexible user- interface, more powerful features and security.

- GSM cordless telephony system to provide a small home base station to work with a standard GSM mobile phone in similar mode to a cordless phone. The base station would be connected to the PSTN.
- Number portability will allow customers to retain their mobile numbers when they change operators or service providers
- Location services to standardize the methods for determining a GSM subscriber's physical location
- Tandem free operation where the compressed speech is passed unchanged over the 64 kbps links between the transcoders, hence improving the voice quality.

3

Wireless Networking and Communication Technology

WIRELESS NETWORKING

Wireless media hubs are Wi-Fi enabled gadgets useful for managing digital photos and videos. You'll find a variety of these products in stores that offer a combination of useful features at different price points. Their primary function is to provide *portable storage*. By plugging in external drives or memory cards into one of hub's ports, you can offload files from phones, tablets or cameras and share them over Wi-Fi with other devices on the network. Some wireless media hubs also offer:

- Wireless access point functionality when connected by Ethernet cable
- The ability to charge the batteries of mobile devices connected to its USB ports

TYPES OF WIRELESS NETWORKS

Wireless PAN

Wireless personal area networks (WPANs) interconnect devices within a relatively small area, that is generally within a person's reach. For example, both Bluetooth radio and invisible infrared light provides a WPAN for interconnecting a headset to a laptop. ZigBee also supports WPAN applications. Wi-Fi PANs are becoming commonplace (2010) as equipment designers start to integrate Wi-Fi into a variety of consumer electronic devices. Intel "My WiFi" and Windows 7 "virtual Wi-Fi" capabilities have made Wi-Fi PANs simpler and easier to set up and configure.

Wireless LAN

A wireless local area network (WLAN) links two or more devices over a short distance using a wireless distribution method, usually providing a connection through an access point for Internet access. The use of spread-spectrum or OFDM technologies may allow users to move around within a local coverage area, and still remain connected to the network. Products using the IEEE 802.11 WLAN standards are marketed under the Wi-Fi brand name. Fixed wireless technology implements point-to-point links between computers or networks at two distant locations, often using dedicated microwave or modulated laser light beams over line of sight paths. It is often used in cities to connect networks in two or more buildings without installing a wired link.

Wireless mesh network

A wireless mesh network is a wireless network made up of radio nodes organized in a mesh topology. Each node forwards messages on behalf of the other nodes. Mesh networks can "self heal", automatically re-routing around a node that has lost power.

Wireless MAN

Wireless metropolitan area networks are a type of wireless network that connects several wireless LANs.

- WiMAX is a type of Wireless MAN and is described by the IEEE 802.16 standard.

Wireless WAN

Wireless wide area networks are wireless networks that typically cover large areas, such as between neighboring towns and cities, or city and suburb. These networks can be used to connect branch offices of business or as a public internet access system. The wireless connections between access points are usually point to point microwave linksusing parabolic dishes on the 2.4 GHz band, rather than omnidirectional antennas used with smaller networks. A typical system contains base station gateways, access points and wireless bridging relays. Other configurations are mesh systems where each access point acts as a relay also. When combined with renewable energy systems such as photo-voltaic solar panels or wind systems they can be stand alone systems.

Cellular network

Fig. Top of a cellular radio tower

A cellular network or mobile network is a radio network distributed over land areas called cells, each served by at least one fixed-location transceiver, known as a cell site or base station. In a cellular network, each cell characteristically uses a different set of radio frequencies from all their immediate neighbouring cells to avoid any interference.

When joined together these cells provide radio coverage over a wide geographic area. This enables a large number of portable transceivers (e.g., mobile phones, pagers, etc.) to communicate with each other and with fixed transceivers and telephones anywhere in the network, via base stations, even if some of the transceivers are moving through more than one cell during transmission.

Although originally intended for cell phones, with the development of smartphones, cellular telephone networks routinely carry data in addition to telephone conversations:

- Global System for Mobile Communications (GSM): The GSM network is divided into three major systems: the switching system, the base station system, and the operation and support system. The cell phone connects to the base system station which then connects to the operation and support station; it then connects to the switching station where the call is transferred to where it needs to go. GSM is the most common standard and is used for a majority of cell phones.
- Personal Communications Service (PCS): PCS is a radio band that can be used by mobile phones in North America and South Asia. Sprint happened to be the first service to set up a PCS.
- D-AMPS: Digital Advanced Mobile Phone Service, an upgraded version of AMPS, is being phased out due to advancement in technology. The newer GSM networks are replacing the older system.

TIME-DIVISION MULTIPLE ACCESS (TDMA)

What is TDMA? TDMA (time division multiple access) is a technology used in digital cellular telephone communication to divide each cellular channel into three time slots in order to increase the amount of data that can be carried.

How it Works? TDMA works by time-division multiplexing: sending multiple signals (each of which has its own time slot) simultaneously on a single carrier in the form of a complex signal, and then recovering the separate signals at the receiving end. For TDMA, the carrier is divided into three time slots, each of which serves one subscriber. The information is broken into tiny data packets, which are transmitted in timed bursts in the 30-megahertz range. At the receiving end, the separate information streams are recovered. See also FDMA (frequency division multiple access) and CDMA (code-division multiple access). TDMA was developed in response to the basic wireless network problem: large numbers of users and limited frequency allotments.

TDMA increases network efficiency by enabling single connections to carry multiple data channels, offering a three-fold increase in capacity over Advanced Mobile Phone Service (AMPS) networks.

Flexible and scalable, TDMA facilitates step-by-step migration to digital operation. TDMA can be implemented seamlessly across both 800- and 1900-MHz networks. Its hierarchical cell structure allows service providers to increase capacity where demand is greatest, in high-use areas.

TDMA is applied in Digital-American Mobile Phone Service, Global System for Mobile communications, and Personal Digital Cellular (PDC). However, each of these systems implements TDMA in a somewhat different and incompatible way. TDMA was first specified as a standard in EIA/TIA Interim Standard 54 (IS-54). IS-136, an evolved version of IS-54, is the United States standard for TDMA for both the cellular (850 MHz) and personal communications services (1.9 GHz) spectrums. TDMA is also used for Digital Enhanced Cordless Telecommunications.

CODE DIVISION MULTIPLE ACCESS (CDMA)

The term CDMA refers to any of several protocols used in so-called second-generation (2G) and third-generation (3G) wireless communications. As the term implies, CDMA is a form of multiplexing, which allows numerous signals to occupy a single transmission channel, optimizing the use of available bandwidth. The technology is used in ultra-high-frequency (UHF) cellular telephone systems in the 800-MHz and 1.9-GHz bands.

CDMA employs analog-to-digital conversion (ADC) in combination with spread spectrum technology. Audio input is first digitized into binary elements. The frequency of the transmitted signal is then made to vary according to a defined pattern (code), so it can be intercepted only by a receiver whose frequency response is programmed with the same code, so it follows exactly along with the transmitter frequency. There are trillions of possible frequency-sequencing codes; this enhances privacy and makes cloning difficult. The CDMA channel is nominally 1.23 MHz wide. CDMA networks use a scheme called soft handoff, which minimizes signal breakup as a handset passes from one cell to another. The combination of digital and spread-spectrum modes supports several times as many signals per unit bandwidth as analog modes. CDMA is compatible with other cellular technologies; this allows for nationwide roaming. The original CDMA standard, also known as CDMA One and still common in cellular telephones in the U.S., offers a transmission speed of only up to 14.4 Kbps in its single channel form and up to 115 Kbps in an eight-channel form. CDMA2000 and wideband CDMA deliver data many times faster.

GLOBAL SYSTEM FOR MOBILE COMMUNICATION

GSM (Global System for Mobile communication) is a digital mobile telephone system that is widely used in Europe and other parts of the world. GSM uses a variation of time division multiple access (Time Division Multiple Access) and is the most widely used of the three digital wireless telephone technologies (TDMA, GSM, and CDMA). GSM digitizes and compresses data, then sends it down a channel with two other streams of user data, each in its own time slot. It operates at either the 900 MHz or 1800 MHz frequency band.

GSM is the de facto wireless telephone standard in Europe. GSM has over 120 million users worldwide and is available in 120 countries, according to the GSM MoU Association. Since many GSM network operators have roaming agreements with foreign operators, users can often continue to use their mobile phones when they travel to other countries.

American Personal Communications (APC), a subsidiary of Sprint, is using GSM as the technology for a broadband personal communications service (personal communications services). The service will ultimately have more than 400 base stations for the palm-sized handsets that are being made by Ericsson, Motorola, and Nokia. The handsets include a phone, a text pager, and an answering machine.

GSM together with other technologies is part of an evolution of wireless mobile telemmunication that includes High-Speed Circuit-Switched Data (High-Speed Circuit-Switched Data), General Packet Radio System (General Packet Radio Services), Enhanced Data GSM Environment (Enhanced Data GSM Environment), and Universal Mobile Telecommunications Service (Universal Mobile Telecommunications System).

ROAMING SERVICE

Roaming service is the ability to get access to the Internet when away from home at the price of a local call or at a charge considerably less than the regular long-distance charges. For example, if you normally get access to the Internet from an access provider in Brooklyn, New York and are travelling to Hong Kong, you can call a designated access provider in Hong Kong. Instead of paying long distance charges to your local provider in Brooklyn, you pay the local phone connection charge in Hong Kong and possibly a modest additional charge for the service.

Roaming service is made possible through Internet service providers (Internet service provider) who have cooperative agreements to grant each others' customers local access to the Internet. Special software allows cooperating ISPs to keep track of and calculate prearranged payments for usage differences. Here's how it works for the user:

- The Internet user must already subscribe to an ISP that offers roaming service arrangements.
- Assuming the ISP does, the user can determine a cooperating ISP in a city to which the user is travelling.
- In the travel location, the user can call the local ISP's designated

phone number through the computer modem, entering information during login that will identify the user's home ISP.

- The "foreign" ISP will contact the ISP and determine that the user is a valid user.
- The "foreign" ISP will grant the user access to the Internet. The user will be able to access e-mail from the home mail server.
- The user will be charged at local phone rates. In addition, depending on the particular service arrangement, the home ISP may levy an additional hourly usage charge of several dollars an hour or a monthly charge in case the service is used during that month.

A similar roaming service is provided by some cooperating cellular telephone or personal digital assistant (personal digital assistant) service providers.

If you are travelling and simply need to be able to exchange e-mail, you can consider getting a freemail membership (usually free) from HotMail, Rocketmail, or other freemail providers. Hotmail also offers POP3 server accounts for access to up to four e-mail accounts you may already have, assuming you remember the POP3 server name and your user IDs and passwords.

If you subscribe to a somewhat global service such as AT&T's WorldNet or the IBM Global Network, you may already be able to access your account in certain cities through your provider's local point-of-presence (POP) on the Internet without having to pay for a long-distance call.

WAP

WAP (Wireless Application Protocol) is a specification for a set of communication protocol to standardize the way that wireless devices, such as cellular telephones and radio transceivers, can be used for Internet access, including e-mail, the World Wide Web, newsgroups, and Internet Relay Chat (Internet Relay Chat). While Internet access has been possible in the past, different manufacturers have used different technologies. In the future, devices and service systems that use WAP will be able to interoperate.

The WAP layers are:

- Wireless Application Environment (WAE)
- Wireless Session Layer (WSL)
- Wireless Transport Layer Security (WTLS)
- Wireless Transport Layer (WTP)
- The WAP was conceived by four companies: Ericsson, Motorola, Nokia, and Unwired Planet (which is now Phone.com).

Wireless Markup Language

WML (Wireless Markup Language), formerly called HDML (Handheld

Devices Markup Language), is a language that allows the text portions of Web pages to be presented on cellular telephone and personal digital assistants (personal digital assistant) via wireless access. WML is part of the Wireless Application Protocol (WAP) that is being proposed by several vendors to standards bodies. The Wireless Application Protocol works on top of standard data link protocols, such as Global System for Mobile communication, code-division multiple access, and Time Division Multiple Access, and provides a complete set of network communication programs comparable to and supportive of the Internet set of protocols.

WML is an open language offered royalty-free. Specifications are available at Phone.com's Web site. According to Phone.com, any programmer with working knowledge of HTML, common gateway interface, and Structured Query Language should be able to write a presentation layer using WML. A filter program can be written or may be available from a vendor that will translate HTML pages into WML pages.

VULNERABILITIES OF MOBILE WIRELESS NETWORKS

The nature of mobile computing environment makes it very vulnerable to an adversary's malicious attacks.

Firstly, the use of wireless links renders the network susceptible to attacks ranging from passive eavesdropping to active interfering as attacks on these links can from any direction and target at any node. This means that a wireless ad-hoc network will not have a clear line of defense, and every node has to be prepared for encounters with an adversary directly or indirectly.

Secondly, mobile nodes are autonomous units that are capable of roaming independently. Since tracking down a particular mobile node in a global scale network cannot be done easily, attacks by compromised node from within the network are more damaging and harder to detect. Third, decision-making in mobile computing environment is sometimes decentralized and some wireless network algorithms rely on the cooperative participation of all nodes and the infrastructure. Furthermore, mobile computing has introduced new type of computational and communication activities that seldom appear in fixed or wired environment. Applications and services in a mobile wireless network can be a weak link as well. To summarize, a mobile wireless network is vulnerable due to its features of open medium, dynamic changing topology, cooperative algorithms, lack of centralized monitoring and management point, and lack of a clear line of defense.

NEED FOR INTRUSION DETECTION

A computer system should provide *confidentiality, integrity* and *assurance* against denial of service. However, due to increased connectivity (especially on the Internet), and the vast spectrum of financial possibilities that are opening up, more and more systems are subject to attack by intruders. These subversion attempts try to exploit flaws in the operating system as well as in application programs and have resulted in spectacular incidents like the Internet Worm incident of 1988.

There are two ways to handle subversion attempts. One way is to prevent subversion itself by building a completely secure system. We could, for example, *require* all users to identify and authenticate themselves; we could protect data by various cryptographic methods and very tight access control mechanisms. However this is not really feasible because:

- In practice, it is not possible to build a completely secure system. Miller gives a compelling report on bugs in popular programs and operating systems that seems to indicate that (a) bug free software is still a dream and (b) no-one seems to want to make the effort to try to develop such software. Apart from the fact that we do not seem to be getting our money's worth when we buy software, there are also security implications when our E-mail software, for example, can be attacked. Designing and implementing a totally secure system is thus an extremely difficult task.
- The vast installed base of systems worldwide guarantees that any transition to a secure system, (if it is ever developed) will be long in coming.
- Cryptographic methods have their own problems. Passwords can be cracked, users can lose their passwords, and entire crypto-systems can be broken.
- Even a truly secure system is vulnerable to abuse by insiders who abuse their privileges.
- It has been seen that that the relationship between the level of access control and user efficiency is an inverse one, which means that the stricter the mechanisms, the lower the efficiency becomes.

The history of security research has taught us a valuable lesson – no matter how many intrusion prevention measures are inserted in a network, there are always some weak links that one could exploit to break in.

We thus see that we are stuck with systems that have vulnerabilities for a while to come. If there are attacks on a system, we would like to detect them as soon as possible (preferably in real-time) and take appropriate action. This is essentially what an Intrusion Detection System (IDS) does. An IDS does not usually take preventive measures when an attack is detected; it is a reactive rather than pro-active agent. It plays the role of an informant rather than a police officer.

BACKGROUND ON INTRUSION DETECTION

In the last three years, the networking revolution has finally come of age. More than ever before, we see that the Internet is changing computing, as we know it. The possibilities and opportunities are limitless; unfortunately, so too are the risks and chances of malicious intrusions. It is very important that the security mechanisms of a system are designed so as to *prevent* unauthorized access to system resources and data. However, completely preventing breaches of security appear, at present, unrealistic. We can, however, try to detect these intrusion attempts so that action may be taken to repair the damage later. This field of research is called

Intrusion Detection

A simple firewall can no longer provide enough security as in the past. Today's corporations are drafting intricate security policies whose enforcement requires the use of multiple systems, both proactive and reactive (and often multi-layered and highly redundant). The premise behind intrusion detection systems is simple: Deploy a set of agents to inspect network traffic and look for the "signatures" of known network attacks. However, the evolution of network computing and the awesome availability of the Internet have complicated this concept somewhat. With the advent of Distributed Denial of Service (DDOS) attacks, which are often launched from hundreds of separate sources, the traffic source no longer provides reliable temporal clues that an attack is in progress. Worse yet, the task of responding to such attacks is further complicated by the diversity of the source systems, and especially by the geographically distributed nature of most attacks.

Intrusion detection techniques while often regarded as grossly experimental, the field of intrusion detection has matured a great deal to the point where it has secured a space in the network defense landscape alongside firewalls and virus protection systems. While the actual implementations tend to be fairly complex, and often proprietary, the concept behind intrusion detection is a surprisingly simple one: Inspect all network activity (both inbound and outbound) and identify suspicious patterns that could be evidence of a network or system attack.

Classification of Intrusion Detection Systems

Intrusions can be divided into 6 main types

1. Attempted break-ins, which are detected by atypical behavior profiles or violations of security constraints.
2. Masquerade attacks, which are detected by atypical behavior profiles or violations of security constraints.
3. Penetration of the security control system, which are detected by monitoring for specific patterns of activity.
4. Leakage, which is detected by atypical use of system resources.
5. Denial of service, which is detected by atypical use of system resources.
6. Malicious use, which is detected by atypical behavior profiles, violations of security constraints, or use of special privileges.

IDS REQUIREMENTS

At least one past effort has identified desirable characteristics for an IDS. Regardless on what mechanisms an IDS is based, it must do the following:

- Run continuously without human supervision,
- Be fault tolerant and survivable,
- Resist subversion,
- Impose minimal overhead,
- Observe deviations from normal behavior
- Be easily tailored to a specific network
- Adapt to changes over time, and
- Be difficult to fool.

We have developed a similar set of requirements along two themes: functional and performance requirements.

Functional Requirements

As the network-computing environment increases in complexity, so do the functional requirements of IDSs. Common functional requirements of an IDS being deployed in current or near-term operational computing environments include the following:

- The IDS must continuously monitor and report intrusions.
- The IDS must supply enough information to repair the system, determine the extent of damage, and establish responsibility for the intrusion.
- The IDS should be modular and configurable as each host and network segment will require their own tests and these tests will need to be continuously upgraded and eventually replaced with new tests.
- Since the IDS is assigned the critical role of monitoring the security state of the network, the IDS itself is a primary target of attack. The IDS must be able to operate in a hostile computing environment and exhibit a high degree of fault-tolerance and allow for graceful degradation.
- The IDS should be adaptive to network topology and configuration changes as computing elements are dynamically added and removed from the network.
- Anomaly detection systems should have a very low false alarm rate. Given the projected increase in network connectivity and traffic, simply decreasing the percentage of overall false alarms may not be sufficient as their absolute number may continue to rise.
- The IDS should be able to learn from past experiences and improve its detection capabilities over time. A self-tuning IDS will be able to learning from false alarms with the guidance of system administrators and eventually on its own.

- The IDS should be able to be easily and frequently updated with attack signatures as new security advisories and security patches become available and new vulnerabilities and attacks are discovered.
- Decision support tools will be necessary to help system administrators respond to various attacks. The IDS will be required not only to detect anomalous events, but also to take automated corrective action.
- The IDS should be able to perform data fusion and be able to process information from multiple and distributed data sources such as firewalls, routers, and switches. As real-time detection demands push networked-based solutions to re-programmable hardware devices that can download new capabilities, the IDS will need to be able to communicate with the hardware-based devices.
- Data reduction tools will be necessary to help the IDS process the information gathered from data fusion techniques. Data mining tools will be helpful in running statistical analysis tools on archived data in support of anomaly detection techniques.
- The IDS should be capable of providing an automated response to suspicious activity.
- Rapid changes in network conditions and limited network administration expertise make it difficult for system administrators to diagnose problems and take corrective action to minimize the damage that intruders can cause.
- The ability to detect and react to distributed and coordinated attacks will become necessary. Coordinated attacks against a network will be able to marshal greater forces and launch many more and varied attacks against a single target. These attacks can be permutations of known attacks, be rapidly evolving, and be launched at little cost to the attackers.
- Distributing the computational load and the diagnostic capabilities to agents scattered throughout the network adds a level of fault-tolerance, but it is often necessary for the system administrator to have control over the IDS from a central location.
- The IDS should be able to work with other Commercial Off-the-Shelf (COTS) security tools, as no vendor toolset is likely to excel in or to provide complete coverage of the detection, diagnosis, and response responsibilities. The IDS framework should be able to integrate various data reduction, forensic, host-based, and network-based security tools. Interoperability and conformance to standards will further increase the value of the IDS.
- IDS data often requires additional analysis to assess any damage to the network after an intrusion has been detected. Although the anomalous event was the first detected, it may not be the first

attempt to gain unauthorized access to the network. Post event analysis will be needed to identify compromised machines before the network can be restored to a safe condition.

- The IDS itself must also be designed with security in mind. For example, the IDS must be able to authenticate the administrator, audit administrator actions, mutually authenticate IDS devices, protect the IDS data, and not create additional vulnerabilities.

Performance Requirements

An IDS that is functionally correct, but that detects attacks too slowly is of little use. Thus we must enumerate several performance requirements for IDSs. The IDS performance requirements include:

- To the extent possible, anomalous events or breaches in security should be detected in real-time and reported immediately to minimize the damage to the network and the loss or corruption of confidential information.
- The IDS must not place undue burden or interfere with the normal operations for which the systems were bought and deployed to begin with. This requirement makes it necessary for the agents to be cognizant of the consumption of network resources for which they are competing.
- The IDS must be scalable. As new computing devices are added to the network, the IDS must be able to handle the additional computational and communication load.

We can divide the techniques of intrusion detection into two main types.

ANOMALY DETECTION

Anomaly detection techniques assume that all intrusive activities are necessarily anomalous. This means that if we could establish a "normal activity profile" for a system, we could, in theory, flag all system states varying from the established profile by statistically significant amounts as intrusion attempts. However, if we consider that the set of intrusive activities only intersects the set of anomalous activities instead of being exactly the same, we find a couple of interesting possibilities:

- Anomalous activities that are not intrusive are flagged as intrusive.
- Intrusive activities that are not anomalous result in false negatives (events are not flagged intrusive, though they actually are).

This is a dangerous problem, and is far more serious than the problem of false positives.

The main issues in anomaly detection systems thus become the selection of threshold levels so that neither of the above 2 problems is unreasonably magnified, and the selection of features to monitor. Anomaly detection systems are also computationally expensive because of the overhead of keeping track of, and possibly updating several system profile metrics. Some systems based on this technique are discussed in Section 4 while a block diagram of a typical anomaly detection system is shown in Figure below.

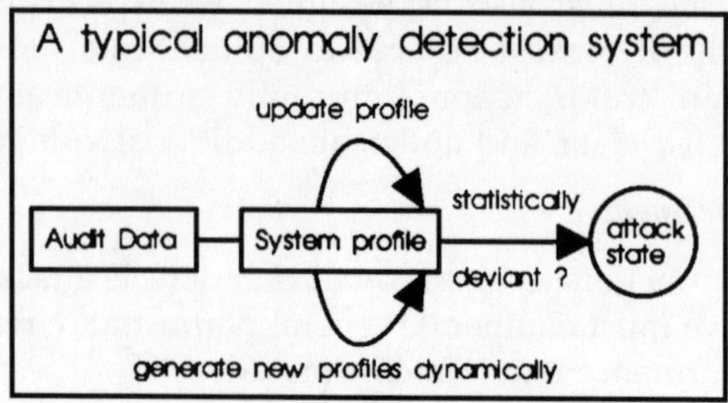

Misuse Detection

The concept behind misuse detection schemes is that there are ways to represent attacks in the form of a pattern or a signature so that even variations of the same attack can be detected. This means that these systems are not unlike virus detection systems — they can detect many or all *known* attack patterns, but they are of little use for as yet unknown attack methods. An interesting point to note is that anomaly detection systems try to detect the complement of "bad" behavior. Misuse detection systems try to recognize known "bad" behavior. The main issues in misuse detection systems are how to write a signature that encompasses *all* possible variations of the pertinent attack, and how to write signatures that do not also match non-intrusive activity. A block diagram of a typical misuse detection system is shown in Figure below.

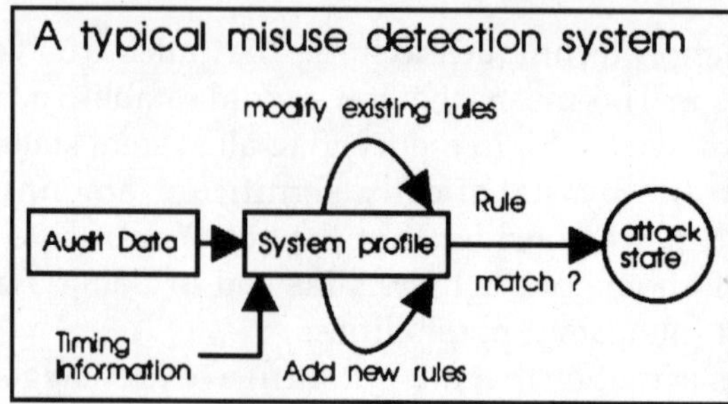

Advantages

- Simplicity and nonintrusiveness (which translate into ease of deployment).

Disadvantages

- Inspecting each packet on the wire is becoming increasingly more difficult with the recent advances in network and wireless technology in terms of complexity and speed.

- Most intrusion detection systems employ a combination of both techniques, and are often deployed on the network, on a specific host, or even on an application within a host.

NETWORK BASED INTRUSION DETECTION:

The most obvious location for an intrusion detection system is right on the segment being monitored. Network-based intrusion detectors insert themselves in the network just like any other device, except they promiscuously examine every packet they see on the wire.

Advantage

- Network-based intrusion detection is straightforward to implement and deploy.

Disadvantage

- Truly shared segments are rare nowadays, which means a single sniffer cannot be relied to monitor an entire subnet. Instead, detection systems must be integrated in the port of Ethernet switches (the ones that have visibility into all packets on the wire), which is not always feasible, even if such a port is available.
- The fact that a single intrusion detection system is servicing the entire segment makes it an easy target for a DoS attack. Such a system should not contain any user accounts other than the privileged (root/Administrator) user; host any unnecessary network services; offer any sort of interactive network access (console access only); or be hosted on an obscure, proprietary operating system.

Host Based Intrusion Detection

While network-based intrusion detectors are straightforward to deploy and maintain, there is a whole class of attacks closely coupled to the target system and extremely hard to fingerprint. These are the ones that exploit vulnerabilities particular to specific operating systems and application suites. Only host-based intrusion detection systems (the ones running as an application on a network-connected host) can correlate the complex array of system-specific parameters that make up the signature of a well-orchestrated attack.

Advantage

- The host-based approach is ideal for those high-availability servers that enterprises rely on for everyday business. The most prevalent advantage of the host-based approach is its ability to detect an inside job-that is, an incident where a lawful user is using local host resources in a manner that violates the company's security policy.

This type of offense would be virtually impossible to unveil with a network-based intrusion detection system; because the user could have console access to the system, his or her actions would not even traverse the wire.

Disadvantage

- Not all is well in the world of host-based intrusion detection, however: Since these systems are closely tied to the operating system, they become yet one more application to maintain and migrate. This is a critical point in an environment where operating system levels are upgraded often, as the intrusion detection system must be kept up to date for it to work efficiently. Also, deploying host-based detectors alone will not protect your enterprise against basic, Network-layer DoS attacks (SYN flooding, ping of death, land attack, and so on). These limitations withstanding, host-based detection should be an integral part of your overall intrusion defense.

ANOMALY DETECTION SYSTEMS

There have been a few approaches to anomaly intrusion detection systems, some of which are described below.

Statistical Approaches

In this method, initially, behavior profiles for subjects are generated. As the system continues running, the anomaly detector constantly generates the variance of the present profile from the original one. We note that, in this case, there may be several measures that affect the behavior profile, like activity measures, CPU time used, number of network connections in a time period, etc. In some systems, the current profile and the previous profile are merged at intervals, but in some other systems profile generation is a one time activity.

Advantages

- The main advantage to statistical systems is that they adaptively learn the behavior of users; they are thus potentially more sensitive than human experts.

Disadvantages

- These statistical approaches can gradually be trained by intruders so that eventually, intrusive events are considered normal, false positives and false negatives are generated depending on whether the threshold is set too low or too high, and relationships between events are missed because of the insensitivity of statistical measures to the order of events.

An open issue with statistical approaches in particular, and anomaly detection systems in general, is the selection of measures to monitor. It is not known exactly what the subset of all possible measures that accurately predicts intrusive activities is. Static methods of determining these measures are sometimes misleading because of the unique features of a particular system. Thus, it seems that a combination of static and dynamic determination of the set of measures should be done. Some problems associated with this technique have been remedied by other methods, including the method involving *Predictive Pattern Generation,* which takes past events into account when analyzing the data.

Predictive Pattern Generation:

This method of intrusion detection tries to predict future events based on the events that have already occurred. Therefore, we could have a rule,

$$E1 - E2 \rightarrow (E3 = 80\%, E4 = 15\%, E5 = 5\%)$$

This would mean that given that events E1 and E2 have occurred, with E2 occurring after E1, there is an 80 per cent probability that event E3 will follow, a 15 per cent chance that event E4 will follow and a 5 per cent probability that event E5 will follow.

Problem

The problem with this is that some intrusion scenarios that are not described by the rules will not be flagged intrusive. Thus, if an event sequence A - B - C exists that is intrusive, but not listed in the rule base, it will be classified as unrecognized.

Solution

The above problem can be partially solved by flagging any unknown events as intrusions (increasing the probability of false positives), or by flagging them as non-intrusive (thus increasing the probability of false negatives). In the normal case, however, an event is flagged intrusive if the left hand side of a rule is matched, but the right hand side is statistically very deviant from the prediction.

Advantages

- There are several advantages to this approach.
- First, rule based sequential patterns can detect anomalous activities that were difficult with traditional methods.
- Second, systems built using this model are highly adaptive to changes. This is because low quality patterns are continuously eliminated, finally leaving the higher quality patterns behind.
- Third, it is easier to detect users who try to train the system during its learning period.

- And fourth, anomalous activities can be detected and reported within seconds of receiving audit events.

Neural Networks

Another approach taken in intrusion detection systems is the use of neural networks. The idea here is to train the neural network to predict a user's next action or command, given the window of 'n' previous actions or commands. The network is trained on a set of representative user commands. After the training period, the network tries to match actual commands with the actual user profile already present in the net. Any incorrectly predicted events actually measure the deviation of the user from the established profile.

Advantages

Some advantages of using neural networks are:

- They cope well with noisy data.
- Their success does not depend on any statistical assumption about the nature of the underlying data.
- They are easier to modify for new user communities.

Disadvantages

However, they have some disadvantages:

- First, a small window will result in false positives while a large window will result in irrelevant data as well as increase the chance of false negatives.
- Second, the net topology is only determined after considerable trial and error.
- The intruder can train the net during its learning phase.

MISUSE DETECTION SYSTEMS:

There has been significant research in misuse detection systems in the recent past.

Expert Systems

These systems are modeled in such a way as to separate the rule matching phase from the action phase. The matching is done according to audit trail events. IDES follows a hybrid intrusion detection technique consisting of a misuse detection component as well as an anomaly detection component. The anomaly detector is based on the statistical approach, and it flags events as intrusive if they are largely deviant from the expected behavior.

To do this, it builds user profiles based on many different criteria (more than 30 criteria, including CPU and I/O usage, commands used, local network activity, system errors etc.). These profiles are updated at periodic intervals. The expert system misuse detection component encodes known intrusion scenarios and attack patterns (bugs in old versions of send mail could be one vulnerability). The rule database can be changed for different systems.

Advantages

- One advantage of the IDES approach is that it has a statistical component as well as an expert system component. This means that the chances of one system catching intrusions missed by the other increase.
- Another advantage is the problem's control reasoning is cleanly separated from the formulation of the solution.

Disadvantages

- There are some drawbacks to the expert system approach too. For example, the expert system has to be formulated by a security professional and thus the system is only as strong as the security personnel who program it. This means that there is a real chance that expert systems can fail to flag intrusions.

It is for the above reason that IDES has an anomaly as well as a misuse detection component. These two components are loosely coupled in the sense that they perform their operations independently for the most part. The IDES system runs on a machine different from the machine(s) to be monitored, which could be unreasonable overhead. Furthermore, additions and deletions of rules from the rule-base must take into account the inter-dependencies between different rules in the rule-base. And there is no recognition of the sequential ordering of data, because the various conditions that make up a rule are not recognized to be ordered.

Keystroke monitoring

This is a very simple technique that monitors keystrokes for attack patterns. Unfortunately the system has several defects — features of shells like *bash, ksh,* and *tcsh* in which user definable aliases are present defeat the technique unless alias expansion and semantic analysis of the commands is taken up. The method also does not analyze the running of a program, only the keystrokes. This means that a malicious program cannot be flagged for intrusive activities. Operating systems do not offer much support for keystroke capturing, so the keystroke monitor should have a hook that analyses keystrokes before sending them on to their intended receiver. An improvement to this would be to monitor system calls by application programs as well, so that an analysis of the program's execution is possible.

Model Based Intrusion Detection

States that certain scenarios are inferred by certain other observable activities. If these activities are monitored, it is possible to find intrusion attempts by looking at activities that infer a certain intrusion scenario. The model-based scheme consists of three important modules. The *anticipator* uses the active models and the scenario models to try to predict the next step in the scenario that is expected to occur. A scenario model is a knowledge base with specifications of intrusion scenarios. The *planner* then translates this hypothesis into a format that shows the behavior, as it would occur in the audit trail. It uses the predicted information to plan what to search for next. The *interpreter* then searches for this data in the audit trail. The system proceeds this way, accumulating more and more evidence for an intrusion attempt until a threshold is crossed; at this point, it signals an intrusion attempt.

This is a very clean approach. Because the planner and the interpreter know what they are searching for at each step, the large amounts of noise present in audit data can be filtered, leading to excellent performance improvements. In addition, the system can predict the attacker's next move based on the intrusion model. These predictions can be used to verify an intrusion hypothesis, to take preventive measures, or to determine what data to look for next.

However, there are some critical issues related to this system. First, patterns for intrusion scenarios must be easily recognized. Second, patterns must always occur in the behavior being looked for. And finally, patterns must be *distinguishing*; they must *not* be associated with any other normal behavior.

IDS ISSUES IN MOBILE ENVIRONMENT

Intrusion detection for traditional, wired networks has been the topic of significant research over the past few years. A problem arises, however, when taking the research for wired networks and directly applying it to wireless networks. Key assumptions are made when designing IDS s for wired networks, such as the difficulty for an attacker to penetrate the physical security of the system, the amount of network bandwidth available to the IDS, etc. Specific problems faced when building IDS for a mobile network are addressed below:

Lack of Physical Wires

The most obvious difference when building an IDS in a wireless environment is the fact that an attacker no longer has to gain physical access to the system in order to compromise the security of the network. Potentially, it is very simple for someone to eavesdrop on network traffic in a wireless environment because they no longer have to break through any physical medium to gain access to the traffic.

Bandwidth Issues

Wireless networks have more constrained bandwidth as compared to wired networks. This problem can manifest itself in a number of different ways when an IDS is using wireless communication to convey information between parts of IDS on separate nodes. An IDS in a mobile environment must be extremely careful to limit the amount of communication that takes place between nodes. A second problem that may possibly arise because of limited bandwidth is erroneous behavior of the IDS due to communication delay between nodes.

Difficulty of Anomaly/Normality Distinction

Distinguishing an anomaly from normalcy has always been somewhat difficult for wired IDS s and wireless IDS s are no different. If nodes in a network receive false or old routing information from a particular node then it is difficult to verify if that particular node has been compromised or not. An attacker could have taken the control of the node to send false information to other nodes in the network, or the node could just be temporarily out of sync due to fast movement or other processing requirements.

Secure Communication Between IDS Agents

It is likely that in a wireless network there will have to be portions of the IDS running on each individual node in the network. Each of these IDS agents will have to communicate with other IDS agents in the network to convey information relating to the status of the system. It is crucial that the information being passed from agent to agent be encrypted as to not allow an attacker to gain access to the communication.

Lack of Centralized Access/Audit Point

The lack of centralized audit points in ad hoc networks present difficult problems for intrusion detection. Most static, wired networks have specific repositories where the IDS can obtain audit data for its misuse and anomaly detection(e.g. switches, routers, gateways, etc.). Without centralized audit points, IDS s on ad hoc networks are limited to use only the current traffic coming in and out of the node as audit data. The algorithms that the IDS uses must be distributed, and take into account the fact that a node can only see a portion of the network traffic.

Possibility of a Node Being Compromised

Since ad hoc networks are dynamic and nodes can move about freely, there is a possibility that one or more nodes could be captured and compromised, especially if the network is in a hostile environment. If the algorithms of the IDS are cooperative, it becomes important to be skeptical of which nodes one can trust. IDS s on ad hoc networks have to be weary of attacks made from nodes in the network itself, not just attacks from outside the network.

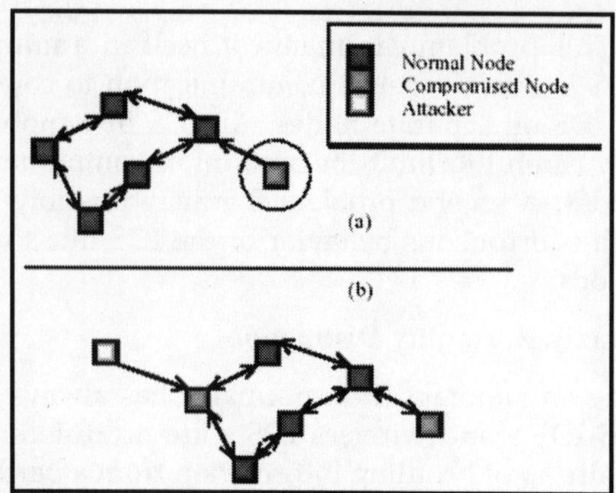

Difficulty In Obtaining Enough Audit Data

Mobile networks do not communicate as frequently as their wired counterparts. Bandwidth issues, and other issues such as battery life, contribute to this factor.

This lack of communication can become a problem for IDS s attempting to define rules of normality for anomaly detection. If only a small amount of data is available to establish normal activity association rules, it is very hard to distinguish an attack from regular network use.

In summary, we must answer the following questions in developing a viable intrusion detection system for mobile ad hoc networks:

- What is a good system architecture for building intrusion detection and response systems that fits the features of mobile ad-hoc networks?
- What are the appropriate audit data sources? How do we detect anomaly based on partial, local audit traces – if they are the only reliable audit source?
- What is a good model of activities in a mobile computing environment that can separate anomaly when under attacks from the normalcy?

NEW ARCHITECTURAL MODELS

It is important to understand that most IDS architectural models are based on static, wired networks. These models alone are insufficient to help design an IDS in a mobile, ad hoc network environment. The architecture addressed is a distributed IDS, where each node on the network will have an IDS agent running on it.

The IDS agents on each node in the network work together via a cooperative intrusion detection algorithm to decide when and how the network is being attacked. The architecture is divided into parts: the Mobile IDS Agents, which reside on each node in the network, and the Stationary

Secure Database, which contains global signatures of known misuse attacks and stores patterns of each users normal activity in a non-hostile environment.

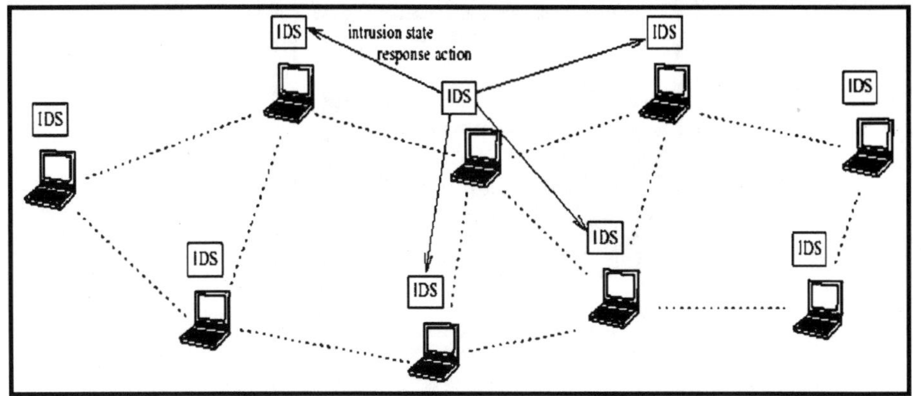

Fig. The IDS Architecture for Wirless Ad-Hoc Network.1

MOBILE IDS AGENTS

Each node in the network will have an IDS agent running on it all times. This agent is responsible for detecting intrusions based on local audit data and participating in cooperative algorithms with other IDS agents to decide if the network is being attacked. Each agent has five parts: the Local Audit Trial, the Local Intrusion Database (LID), the Secure Communication Module, the Anomaly Detection Modules (ADM s), and the Misuse Detection Modules (MDM s).

The Local Audit Trial

Each agent must constantly check the audit data to decide that an intrusion is not taking place. The Local Audit Trial will consist of specific items out of the network traffic as well as user commands to the node. The Local Audit Trial is responsible for selecting only the items it needs out of the network traffic and system audit data in order to minimize the size of the audit data collected.

A audit data is collected by the Local Audit Trial, it is passed to the Misuse Detection Modules and the Anomaly Detection Modules for further analysis. The Local Audit Trial is only responsible for gathering and storing audit data, not processing it.

The Local Intrusion Database (LID)

The LID is a local database that warehouses all information necessary for the IDS agent, such as the signature files of known attacks, the established patterns of users on the network, and the normal traffic flow of the network. The Anomaly Detection Modules and Misuse Detection Modules communicate directly with the LID to determine if an intrusion is taking place.

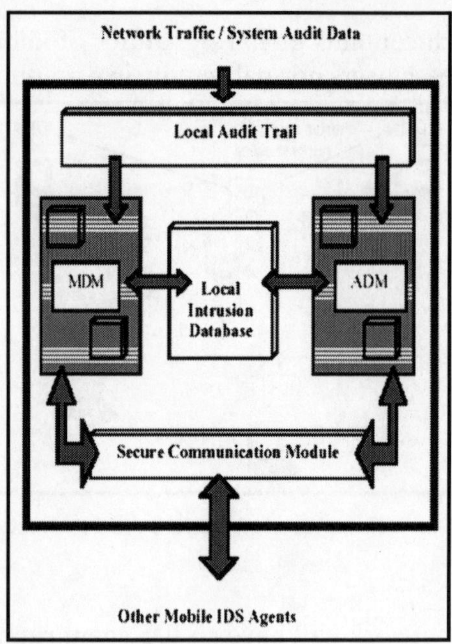

Fig. A Graphical Representation of a Proposed Mobile IDS Agent.

The Secure Communication Module

The Secure Communication Module is necessary to enable an IDS agent to communicate with other IDS agents on other nodes. It will allow the MDM s and ADM s to use cooperative algorithms to detect intrusions. It may also be used to initiate a global response when an IDS agent or a group of IDS agents detects an intrusion. Basically, any communication that needs to occur from one IDS agent to another will use the Secure Communication Module.

Data communicated via the Secure Communication Module will need to be encrypted in order to ensure that the data received by an IDS agent is accurate and has not been tampered with. The Secure Communication module is only used by IDS agents and does not communicate any other type of information between nodes. It must share the bandwidth that the mobile device uses for normal data transmission, so it is required to be efficient, and can only use the amount of bandwidth in needs.

Also, the Secure Communication module must process information coming to the IDS agent from other agents in the network. For this reason, it must be fast and efficient, so as not to take away from the processing time of the mobile unit.

The Anomaly Detection Modules (ADM s)

Each Anomaly Detection Module is responsible for detecting a different type of anomaly. There can be from one to many Anomaly Detection Modules on each mobile IDS agent, each working separately or cooperatively with other ADM s. For example, one ADM might be looking for strange network traffic patterns, while another ADM might be watching user input speed.

If an ADM can identify an anomaly based solely on the data in the Local Intrusion database, then it can initiate a local and global response to the intrusion. An example of a local response could be to shut down the node, rendering it useless to an attacker. A possible global response would be to use the Secure Communication Module to alert other IDS agents, allowing them to reconstitute a network while excluding the compromised node.

If the amount of data in the Local Intrusion database is not sufficient to determine if the present activity should be classified as an intrusion, then it is possible for the ADM to use the Secure Communication module to query other nodes in the network to get help in identifying an intrusion.

The Misuse Detection Modules (MDM s)

The Misuse Detection Modules functions similarly to the ADM s on the IDS agent. The primary difference is that MDM s only identify known patterns of attacks that are specified in the Local Intrusion Database. Like the ADM s, if the audit data available locally is enough to determine if an intrusion is taking place, the proper response can be initiated. It is also possible for a MDM to use a cooperative algorithm to identify an intrusion. If a MDM needs more information from other IDS agents on other nodes, it would be expected to use the Secure Communication module to interact with them. Using the information given by other IDS agents, the MDM might be able to predict an intrusion with more accuracy.

STATIONARY SECURE DATABASE

The Stationary Secure Database (SSD) in this architecture acts as a secure, trusted repository for mobile nodes to obtain information about the latest misuse signatures and to find the latest patterns of normal user activity. It is assumed that the attacker will not compromise the Stationary Secure Database, as it stored in area of high security. To ensure that the SSD will not be compromised it is kept stationary and not placed in a hostile environment where attacker attack is likely. It is also assumed that no physically compromised node will come in contact with the SSD, since the attacker will not be given physical access to the area where the SSD resides. Although these are severe restrictions, they can be accommodated through operational procedures and physical security.

The mobile IDS agents will collect and store audit data (such as user commands, network traffic, etc.) while in the field, and will transfer this information when it is attached to the SSD. The SSD will then use this information for data mining of new anomaly association rules. The use of the SSD to mine new anomaly rules is beneficial to the IDS for three reasons. First, the SSD will be fixed, fast machine that is capable of mining rules much faster than on slower, mobile nodes. Secondly, the processing time used to mine the new rules of anomaly will not take away from the processing time of the mobile nodes. The SSD puts the task of creating new rules for anomaly detection on the wired server and away from the mobile nodes. And thirdly, the SSD is capable of having much more storage capacity to store an abundance of audit data collected from the nodes. It is very likely that the mobile nodes will not have enough storage to store substantial amounts of audit data, but by uploading audit data to the SSD, no data is deleted because of lack of storage space.

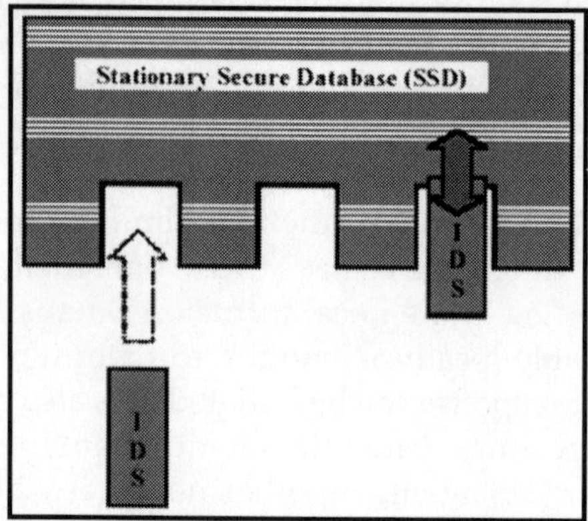

Fig. Mobile IDS Agents Interacting with SSD

The SSD will also be the place where the system administrator can specify the newest misuse signatures. When the IDS agents are connected to SSD, they will gain access to the latest attack signatures automatically. This will make it much easier to update all the nodes in the network to keep up with the latest attacks. Instead of manually updating the attack files in the Local Intrusion Database of each individual node, or using the Secure Communication device on each node to communicate the new signatures, the SSD will be responsible for communicating the new attack signatures to each individual IDS agent.

One of the best reasons for using the SSD to communicate the new attack signatures, and establish new patterns of normalcy, is to limit the amount of communication that must take place between IDS agents in the mobile ad hoc network. As stated earlier, the IDS agents should not use very much bandwidth, because it is limited and in use by other applications on the mobile node. The use of SSD allows the IDS agents to not continually have to share information in order to update their Local Intrusion Database. Communication between the SSD and the IDS agents will be very quick and efficient, as there should be no threat of attack. By relying on the SSD to be a trusted source of update information, the IDS agent no longer has to use cooperative algorithms to determine if the information being sent is trustworthy or not.

It is feasible that the SSD have other functions besides updating the intrusion detection information on each IDS agent, which could be a place where the mobile nodes could charge their batteries while receiving the latest IDS information for example. This way the time spent at the SSD would be more efficient and not be used for just intrusion detection. Despite the benefits of having a SSD in a mobile IDS architecture, there are few disadvantages of relying on a stationary database to provide vital IDS information. If a SSD is used, mobile nodes will have to be attached to the non-mobile database periodically to stay up-to-date with the latest intrusion information. This may not be an option for some mobile, ad hoc environments. Also, since the SSD must be a trusted source, it cannot be taken onsite without significant risk. If a mobile IDS agent detects a new intrusion while in a hostile environment, it cannot be attached to the SSD in order to communicate the new attack patterns. And, even if it could, other nodes would be in the hostile environment and would not be able to attach to the SSD right away to get the new signature. However, these problems might be solved if the IDS agents can communicate the new patterns of attack to each other via the Secure Communication module while in the hostile environment. This way, only information that has to be communicated right away will be sent over the wireless channels, and the less time sensitive information can be gathered later at the SSD when time permits.

ANOMALY DETECTION IN WIRE-LESS AD-HOC NETWORKS

In this section we discuss how to build an anomaly detection models for wireless networks. Detection based on activities in different network layers may differ in the format and the amount of available audit data as well as the modeling algorithms.

BUILDING AN ANOMALY DETECTION MODEL

Framework

The basic premise for anomaly detection is that there is intrinsic and observable characteristic of normal behavior that is distinct from that of abnormal behavior. Entropy and conditional entropy are used to describe the characteristics of the normal information flows and use the classification algorithms to build anomaly detection models. We can use a classifier trained using normal data to predict what is normally the next event given the previous n events. In monitoring when the actual event is not what the classifier has predicted there is an anomaly. When constructing a classifier features with high information gain are needed.

Using this frame work we employ the following the procedure for the anomaly detection.

- Select or partition audit data so that the normal data set has *low Entropy*

- Perform appropriate data transformation according to entropy measures
- Compute classifier using training data.
- Apply the classifier to test it.
- Post process alarms to produce intrusion reports.

Attack Models

Route logic compromise: This type of attacks behaves by manipulating routing information, either externally by parsing false route messages or internally by maliciously changing routing cache information. In particular, we consider several special cases:

- *Misrouting:* Forwarding a packet to an incorrect node; and
- *False message propagation:* Distributing a false route update.

Traffic pattern distortion: This type of attacks changes default/normal traffic behavior:

- Packet dropping;
- Packet generation with faked source address;
- Corruption on packet contents; and
- Denial-of-service.

Audit Data: The two local data sources used for anomaly detection:

1. Local routing information, including cache entries and traffic statistics and
2. Position locator, or GPS, which we assume will not be compromised and can therefore reliably provide location and velocity information of nodes within the whole neighborhood.

We use only local information because remote nodes can be compromised and their data cannot be trusted.

Feature Selection: Feature selection is a critical step in building a detection model. specifically, since we use classifiers as detectors, we need to select and/or construct features, from the available audit data, that have high information gain. The criteria of information gain is not a priori. We use an unsupervised method to construct the feature set. First, we constructed a large feature set to cover a wide range of behaviors. It is not efficient to run all experiments with all of these features. A small number of training runs can be conducted with the whole set of features on small audit data traces randomly chosen from a previously stored audit logs. For each training run, a corresponding model is built. The features that appear in the models and has weights not smaller than a minimum threshold are selected into the essential feature set. For different routing protocols and different scenarios, the essential feature set is different.

In practice, we expect the feature set needs to be updated after certain period, as the characteristics of routing behavior can change with time. Classifier: We use two classifiers in our study. One is a decision-tree equivalent classifier, RIPPER, a rule induction program. The other is a Support Vector Machine classifier, SVM Light. RIPPER is a typical classifier in that it searches the given feature space and computes rules that separate data into appropriate (intended) classes. SVM Light instead pre-processes the data to represent patterns in much higher dimension than the given feature space. The heuristic is that with sufficiently high dimension, a hyper plane, thus achieving the goal of classification, can separate data. SVM Light can produce a more accurate classifier than RIPPER when there are underlying complex patterns in the data that are not readily represented by the given set of features.

Post Processing: Given an execution trace, we first apply a detector to examine each observation. Then a post-processing scheme is used to examine the predictions and generate intrusion reports. Areas where anomaly detection can be used:

The two main areas where we need anomaly detection is ad-hoc networks is

- Abnormal Updates to the routing table.
- Abnormal activities in other layers.

Abnormal Updates in the routing tables

The two most important factors that are required for the anomaly detection are Low False positive rate (percentage of normalancy variations detected as anomalies) High true positive rate (percentage of anomalies detected)

A routing table usually contains, at the minimum the next hop to each destination node and the number of hops. The physical movement of nodes or network membership changes causes a legitimate movement in the routing table. Our objective in this study is to lead a better understanding of the important and challenging issues in intrusion detection for ad-hoc routing protocols. First using a given set of training, testing and evaluation scenarios, and modeling algorithms, we can identify which routing protocol, with potentially all its routing information used, can result in better performing detection models.This will help Answer the question "what information should be included in the routing table to make "intrusion detection effective". This finding can be used in designing more robust protocols.

Next, using a given routing protocol, we can explore the feature space and algorithm space to find the best performing model. This will give the insight to the general practices of building intrusion detection for wireless networks.

Abnormal Activities In other layers

At the wireless application layer, the trace data can use the service as the class (i.e., one class for each service), and can contain the following features: for the past s seconds, the total number of requests to the same service, the number of different services requested, the average duration of the service, the number of nodes that requested (any) service, the total number of service errors, etc. A classifier on the trace data then describes for each service the normal behaviors of its requests. Many attacks generate different statistical patterns than normal requests. Since the features described above are designed to capture the statistical behavior of the requests, the attacks, when examined using the feature values, will have large deviations than the normal requests.

IMPLEMENTED APPROACHES

Following are the some of the intrusion detection techniques used in wireless and ad hoc networks.

IEEE 802.11

The IEEE 802.11 standard provides several mechanisms intended to provide a secure operating environment. The IEEE 802.11 standard defines the physical layers and the MAC sub layers for the wireless LANs. There are three different physical layers.

- Frequency hopping Spread Spectrum Radio.
- Direct sequence spread spectrum Radio.
- Base band infrared.

The MAC layer is common for all these layers. The IEEE 802.11 defines two authentication schemes:

Open System Authentication:

Open system authentication is the default authentication protocol for 802.11. As the name implies, open system authentication authenticates anyone who requests authentication. A terminal announces that it wishes to associate with an access point, and typically the access point allows the association. Essentially it provides NULL authentication process.

Shared Key Authentication

Shared key authentication uses a standard challenge and response along with a shared secret key to provide authentication. The shared key Authentication requires that the Wired Equivalent privacy protocol (WEP) Algorithm be implemented on both the wireless terminal and the access point. The station wishing to authenticate, *the initiator,* sends an authentication request management frame indicating that they wish to use "shared key" authentication.

The recipient of the authentication request, *the responder,* responds by sending an authentication management frame containing challenge text to the initiator. The challenge text is generated by using the WEP pseudo-random number generator (PRNG) with the "shared secret" and a random initialization vector (IV)[2]. Once the initiator receives the management frame from the responder, they copy the contents of the challenge text into a new management frame body. This new management frame body is then encrypted with WEP using the "shared secret" along with a new IV selected by the initiator. The encrypted management frame is then sent to the responder.

The responder decrypts the received frame and verifies that the 32-bit CRC integrity check value (ICV) is valid, and that the challenge text matches that sent in the first message. If they do, then authentication is successful. If the authentication is successful, then the initiator and the responder switch roles and repeat the process to ensure mutual Authentication.

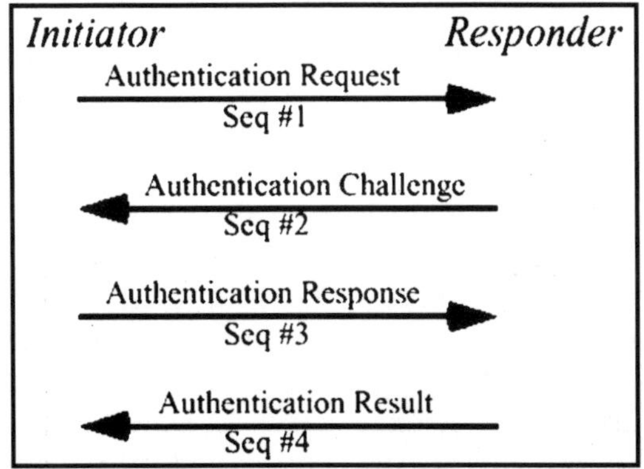

Fig. Mutual Station Authentication Using Shared Keys.

Mobiles that are allowed to connect to the network use the same shared key, so this authentication method is only able to verify if the particular mobile belongs to the group allowed to connect to the network, but there is no way to distinguish one mobile from another. Also there are no means available to authenticate the network.

The IEEE 802.11 does not define any key management functions. The IEEE 802.11 defines an optional WEP mechanism to implement the confidentiality and integrity of the traffic in the network.

WEP is used at the station-to-station level and does not offer any end-to-end security. Using, say, the playback attack, could easily fool the Shared Key Authentication scheme. Hence, anyway an additional authentication mechanism is needed.

Secure key generation and distribution:

The mobile systems have constraints like minimal computational capabilities and authentication and the Secure key generation and distribution capability is required by any system, which contains cryptographic authentication, confidentiality and identification. Developing faster and more powerful hardware components, which require less Energy and changing the algorithmic and protocol design of the current system would be useful to meet the future needs.

Current Approaches for the Key generation:

- This requires the manufacturers and Service provider to develop a special distribution channel.
- Security of keys should be ensured from the time the keys are sent to the Service provider. from the manufacturer.
- This approach is unacceptable to both the Service provider and the manufacturer.

Over-the-air phone activation with key exchange

Over-the-air phone is the most preferred approach and requires a collaborative key generation and distribution between the mobile unit and the Service provider. The current over-the-air service provisioning (OTASP) uses the Diffie-Hellman key exchange between the Service provider and mobile unit to exchange a symmetric key called A-key (Authentication Key). This is used as the "master" symmetric key, which is then used to generate the session keys; these in turn are used for authentication and encryption. This method is inefficient because of the amount of time it takes to set up an authentication session. Also they do not consider exact notions of cryptographic security. The key exchanged needs to be chosen by the mobile unit only. Less efficient due to constrained computational environment of mobile agent.

Over-the-air phone Activation Approach

Secure implementation

Service Provider broadcasts public information together with the signature of the C.A. (Certification Authority whose digital signature will certify the validity of a S.P.) on that information. Mobile Unit Authenticates the Service Provider using the built-in C.A. public key. Generate a random session key SK and a random authentication pad AP. Encrypt (SK, AP) with Service Providers public key and send it to Service provider.

Service Provider decrypts the session key and authenticated pad sent by Mobile unit. Generate a key (A-key) and encrypt it together with the authentication pad AP using a secure symmetric encryption algorithm with the received session key SK. Mobile unit decrypts the A-key and verify that its last bits are equal to the AP.

To make above system efficient for the practical purposes, we need to find an efficient way to authenticate the Service Provider to the mobile agent and we need to find a efficient public-key encryption method to transmit the session key fro mobile unit to service provider.

Semantic Security ensures that no partial information about the session key can be obtained, given the encryption and even if the attacker managed to guess the session key sent by the mobile unit, he would be unable to verify it. *Plain text* awareness requires that the attacker not only cannot obtain any advantage from choosing a random message, but he is required to know the whole plain text.

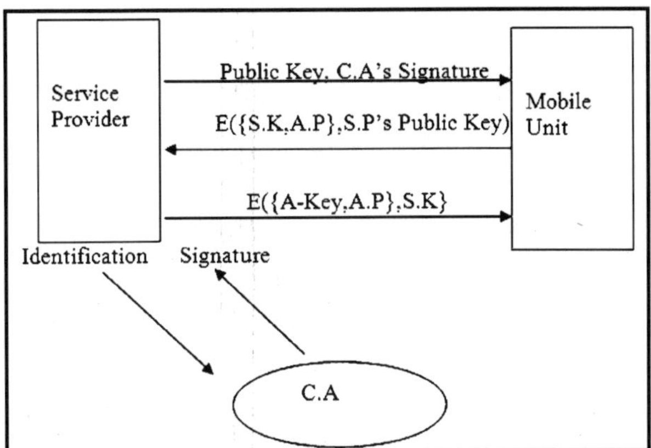

Secure Symmetric encryption

The notion of security for symmetric encryption schemes is also semantic security together with plain text awareness. For security we want to guarantee that even if an attacker intercepts the encryption and manages to guess either the SK or the pad AP (obtains a candidate A-key by decrypting the cipher text) it is still impossible for him to verify that the A-key he obtained is the one included in the cipher text at the same time M.U. is able to verify that the key sent was generated by someone who knew both the AP and SK, i.e., it is not simply a random number.

This is seemingly a contradicting goal, since it requires semantic security and plaintext awareness. But plaintext awareness requires some form of authentication of the plaintext, and here we want to prevent an attacker from verifying such authentication on the plaintext even if he guesses the encryption key. The solution is to separate the session key from the authentication pad, thus M.U. can authenticate the encrypted string using AP but the attacker cannot verify whether his guess of a (SK, AP) pair is correct since he does not know the relation between SK and AP.

We have seen that the solution proposed for Secure key generation and distribution is more efficient and secure than the current practices and can be the first step in achieving cryptographically provable security in an environment with tight restrictions on computational power.

Mitigating Routing Misbehavior

In this section we present the watchdog and the pathrater tools for detecting and mitigating routing misbehavior. We also describe the limitations of these methods. Watchdog detects misbehaving nodes. Suppose there exists a path from node S to D through intermediate nodes A, B and C. Node A cannot transmit all the way to node C, but it can listen in on node B's traffic. Thus, when A transmits a packet for B to forward to C, A can often tell if B transmits the packet. If encryption is not performed separately for each link, which can be expensive, then A can also tell if B has tampered with the payload or the header.

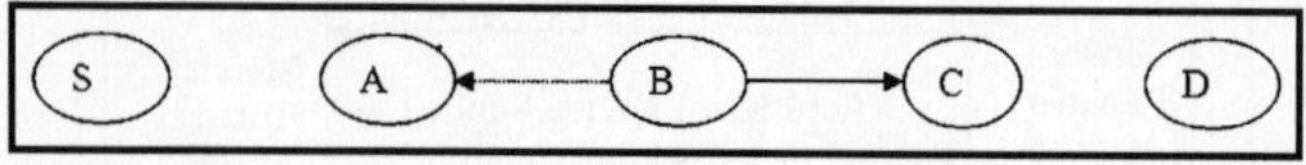

Fig. Watch Dog Operation

Every time a node fails to forward the packet, the watchdog increments the failure tally. If the tally exceeds a certain threshold bandwidth, it determines that the node is

Misbehaving. The watchdog technique has advantages and weaknesses. DSR with the watchdog has the advantage that it can detect misbehavior at the forwarding level and not just the link level. Disadvantages with this technique is that it might not detect a misbehaving go of a node in the presence of:

- Ambiguous Collisions
- Receiver Collisions
- Limited Transmission Power
- False Misbehavior
- Partial Dropping

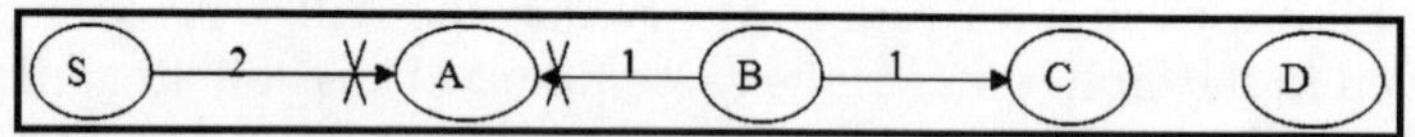

Fig. Node A does not hear the B forwarding packet 1 to C,because B's transmission collides at A with packet 2 from source S.

Pathrater: Knowledge of misbehaving nodes with link reliability data to pick the route most likely to be reliable. Each node maintains a rating for every other node it knows about in the network. It calculates a path metric by averaging the node ratings in the path. We choose this metric because it gives a comparison of the overall reliability of different paths and allows pathrater to emulate the shortest length path algorithm when no reliability information has been collected, as explained below. If there are multiple paths to the same destination, we choose the path with the highest metric. Note that this differs from standard DSR, which chooses the shortest path in the route cache. Further note that since the pathrater depends on knowing the exact path a packet has traversed, it must be implemented on top of a source routing protocol.

THE BASICS OF WIRELESS COMMUNICATION TECHNOLOGY

RADIOFREQUENCY AND THE ELECTROMAGNETIC SPECTRUM

All energy emitted as waves is called radiation. These waves have different lengths, and are typically classified by frequency; longer wavelengths are lower frequency, while shorter wavelengths are higher frequency. The frequency of a wave, measured in hertz (Hz), determines its location along the electromagnetic spectrum. Visible light, for instance, is located approximately in the middle of the spectrum, and acts as the dividing line between non-ionizing and ionizing radiation. Ionizing radiation is high-frequency radiation, which produces enough energy to alter the atomic structure of cells. Antennae used for wireless communication emit radiofrequency radiation, which is low-frequency, non-ionizing radiation, incapable of changing atomic structures.

CELLULAR COMMUNICATION

In 1981, the FCC established rules and procedures for developing a wireless cellular phone system. The FCC licensing rules created 734 cellular markets within the United States, its territories and possessions, and set aside 50 MHZ of spectrum in the 800 MHZ frequency band for two competing cellular systems in each market. Licensees in the top thirty markets were selected using comparative hearings, while those in the remaining markets were determined by lottery. Each licensed cellular carrier operates its own system, utilizing its designated 25 MHZ of spectrum, consisting of 395 voice channels and 21 control channels.

A cellular system is made up of numerous, contiguous, individual cells, each covering a specific broadcast area ranging from one urban block or less to ten square miles or more. Each cell has its own base station, consisting of antennae and associated electronic equipment, which sends and receives information to and from mobile phones. A typical base station utilizes either several *omnidirectional* antennas that look like poles or whips, [ten] to [fifteen] feet in length, or a number of *sector* antennas that look like rectangular panels. The dimensions of a sector antenna are typically [one] foot by [four] feet. Base station antennae may be supported by towers (either lattice type or monopole, and typically 50 to 200 feet high), or may be mounted on tall structures such as water towers or rooftops. Height is a crucial factor in siting and designing these facilities, since wireless communication requires a clear line-of-sight for effective transmission and reception.

Cellular systems are generally analog systems, which can accommodate one user per frequency (or channel), per cell. Each cell is allocated a specific set of the system's channels, with adjacent cells assigned a different set of channels in order to avoid interference. A base station uses its set of channels to relay calls to and from mobile phones, and constantly monitors the strength of the signal between it and each individual mobile phone. As a mobile phone user drives from one cell to another and the signal between the phone and the base station weakens, his call is automatically handed off to a base station nearer to him, and his call is unnoticeably switched to a frequency assigned to that cell. Each cell is also connected via land line or microwave to a switching office, which in turn is connected to the regular land line public telephone network. This combination of radiofrequency, microwave and land line connections provides the link between mobile phone users and land line telephone network users.

PERSONAL COMMUNICATIONS SERVICES

Personal Communications Services (PCS) represents the state of the art in wireless communication. While cellular's analog technology accommodates only one user per frequency at any given time, PCS systems are digital and can accommodate multiple users of a single frequency simultaneously. Additionally, unlike analog cellular systems which can transmit only voice, digital PCS systems are capable of transmitting voice, data and video transmissions. PCS is a broad term describing a number of systems and services which enable a user to communicate anytime, anywhere, and in any form.... The basic idea is based on the notion of allocating a single ID called a Universal Personal Telecommunication (UPT) number to a device.... analogous to a Social Security number for communication!

PCS systems operate independently of traditional cellular systems, and are governed by a separate set of FCC licensing rules. The FCC designated two types of PCS service areas: 51 Major Trading Areas (MTAs), each approximately the size of a mid-sized state, and 493 Basic Trading Areas (BTAs), each the size of several counties. These service areas do not correspond to, nor do they necessarily overlap, the traditional cellular markets. The FCC also allocated a total of 120 MHZ of spectrum for PCS systems, which it divided into three 30 MHZ blocks (blocks A through C) and three 10 MHZ blocks (blocks D through F). The [PCS] licenses in frequency blocks A and B are awarded on an MTA basis; the licenses in frequency blocks C, D, E, and F are awarded on a BTA basis.

PCS technology requires a cell system of facilities structured very much like traditional cellular systems, since both technologies are premised on the idea of reusing frequencies over and over again in cells throughout their systems, although PCS systems operate on a much higher frequency band (1850-2200 MHZ). The main differences between traditional cellular systems and PCS systems stem from PCS' use of microcellular and digital technology. PCS microcells, as the name implies, are much smaller than traditional cells. While regular cellular telephone systems typically locate cells eight miles apart, some microcells, particularly those in urban areas, may be as small as a city block or the floor of an office building. PCS systems, then, are made up of a larger number of smaller cells.

PCS systems are also more cost effective than cellular systems. Installation costs for PCS systems average around $5,000 per cell, while an average cellular system cell installation may cost anywhere from $50,000 to one million dollars. In addition to lower average installation costs per cell, costs to consumers are generally lower, since [t]he cost of using PCS can be shared by a larger number of consumers because PCS systems will have the capability of serving almost twenty times as many people as cellular systems. The cost benefits of PCS systems, both to carriers and consumers, coupled with PCS' facilitation of increased mobility and service options, indicate that PCS is the current ideal in wireless communication.

TECHNOLOGICAL TRENDS IN WIRELESS TELECOMMUNICATIONS

Hatfield Associates, Inc. (HAI) entered into an agreement with Gallaudet University (Gallaudet). Under the terms of the agreement, HAI was engaged to provide certain consulting services in support of a project entitled "Universal Telecommunications Access" being conducted by Gallaudet on behalf of the National Institute of Disability Rehabilitation Research of the U.S. Department of Education. Among other things, HAI was tasked with preparing a document, including a technological forecast, on the development of mobile or "wireless telecommunications." This report constitutes that document and conveys the results of HAI's study carried out under that portion of the consulting agreement.

The purpose of this report is to provide readers with a basic introduction to wireless telecommunications, its underlying technological trends, and the associated regulatory framework and industry structure. The goal is to integrate this information on wireless telecommunications with information about (a) issues surrounding the accessibility and usability of telecommunications to people with disabilities and (b) how rapidly evolving wireless technologies can be particularly useful to this same community. The primary focus of the report is on terrestrial- and satellite-based land mobile radio services as opposed, for example, to maritime or aeronautical radio services. It is primarily aimed at readers without a strong technical background and to provide those readers with a better understanding of the implications of the developments in wireless telecommunications on persons with disabilities.

THE SIGNIFICANCE OF WIRELESS TELECOMMUNICATIONS TO PEOPLE WITH DISABILITIES

Wireless telecommunications holds particular promise for people with disabilities because it enhances both mobility and communications, two functions that are often challenging for people with certain kinds of disabilities. Ordinary cordless telephones have long been useful devices for people who have mobility disabilities and cannot rush to the telephone. Similarly, cellular telephones have been valuable safety devices for people with mobility disabilities traveling alone, and they can help compensate for the lack of accessibility of many pay telephones. Alphanumeric pagers and other wireless data communications systems have been used for communicating with deaf employees who are in the field and who otherwise may not have been able to hold jobs that required frequent communication with a dispatcher or other mobile employees.

The long-term trends in wireless telecommunications hold out even more promise for the future. One concept, associated with the notion of Personal Communications Services, involves each customer having a single, unique telephone number or network address rather than separate ones for home telephone, office telephone, cellular telephone, fax machine, and so on. This concept is appealing to the disabled as it is to the general public. Likewise, the concept of including all modes of electronic communications — voice, data, image (graphics) and video — in a single interface has considerable appeal to the disabled as it does to the public more broadly. Indeed, the ability to communicate anytime, anywhere, in any mode, coupled with the power of intelligent, programmable networks and end user equipment, will create a potent platform upon which to serve disabled subscribers. If the technology and marketplace support this vision of the future, wireless will revolutionize telecommunications — not only for the general public, but for disabled Americans as well.

As long as new and/or advanced technologies are limited or specialized, access to them is less likely to be critical to an individual participating fully in society. But when technologies become pervasive, rather than limited or specialized, concerns over their accessibility and usability escalate. Without access to these broadly available and essential new capabilities, people with disabilities can become isolated rather than empowered. Examples of this abound. When the first telephones were introduced, they were probably not of great concern to people who were deaf. However, when the telephone came to dominate personal and commercial communications, the effect was devastating. Gaining access to early, text based, computer systems was relatively straightforward for blind people. Therefore, they enjoyed enhanced access to print, communications and to new opportunities for employment. Then, when the graphical user interface became the office standard for personal computers, it severely threatened all of this progress.

In the 1980s and 1990s, the use of wireless telecommunications increased dramatically. For instance, prior to 1984, there were only a few thousand mobile telephone subscribers in the United States. However, with the development, licensing and construction of cellular mobile radio systems, the number grew to 44 million by the end of 1996. Similarly, in the same period, the number of radio paging ("beeper") subscribers increased to 42 million. As the number of users has increased, the problems of accessibility to wireless telecommunications for people with certain kinds of disabilities have raised concern. For example, most analog cellular telephones are not hearing-aid compatible (HAC) and their acoustic output is often lower than that of an ordinary wireline telephone. The fact that wireless telephones often do not couple well to the ear sometimes exacerbates this problem of low acoustic output, causing further losses of acoustic energy. Moreover, most cellular telephones do not couple acoustically to text telephones (TTY) widely used by the deaf and many do not have jacks to allow direct connection to TTY devices.

In current-generation digital cellular telephones, the situation is even more serious. Like all two-way wireless communications devices, these telephones radiate (transmit) electromagnetic fields — otherwise they would be unable to fulfill their intended purpose. However, in many implementations of this advanced digital technology, these transmissions occur as a regular series of bursts that are heard as a buzz in nearby hearing aids. These same electromagnetic fields can also render useless the telecoils that are placed in hearing aids to allow users to access ordinary telephone handsets. Depending on the type of hearing aid worn by the user and other factors, this interference may be picked up several feet away from the cellular telephone that is in use. Furthermore, and somewhat ironically, because these new digital cellular systems are primarily optimized for the transmission of the human voice, they cannot be used to transmit TTY signals.

Thus, to summarize, the rapid evolution of wireless telecommunications systems and devices is significant to people with disabilities for at least two reasons. On the one hand, the evolution (some would say revolution) holds out particular promise because it can enhance both mobility and communications, two functions that are often challenging for people with disabilities. On the other hand, experience has shown that, if these wireless systems are not designed, developed, and fabricated to be accessible to — and usable by — individuals with disabilities, then, as they become more pervasive, people with disabilities will become isolated rather than empowered.

Organization of the Report

The balance of this report is divided as follows: Section, I.D., immediately below, provides a brief introduction to the regulatory framework and strucuture of the wireless telecommunications industry in the United States. Section II provides an introduction to the underlying principles of wireless communications and briefly describes the types of services that have traditionally been provided on land mobile radio systems. It is intended to provide readers with the necessary technical background to understand the balance of the report. Section III provides background on the types of land mobile radio systems (broadly defined) that are currently in use. In essence, it reviews the capabilities of existing systems and describes their limitations in general terms (i.e., without special regard to issues associated with their use by the people with disabilities). Section IV describes new and evolving wireless telecommunications systems that are being developed and deployed in order to overcome the limitations identified in Section III. The focus of the section is on technological and service trends in the medium term – i.e., developments that can accurately be foreseen based upon current developments in infrastructure and end user systems and equipment. Since many of the system-level developments are being driven by other advancements (e.g., in digital integrated circuits – "chips" – and radio frequency (RF) devices), the section begins with a review of such advancements. Next, Section V provides a brief review of recent legislative and regulatory developments that will influence the future development of wireless systems and services. This reflects the notion that, while the primary emphasis of the entire report is on technological developments, those developments will be strongly influenced by policies and regulations relating to the management of the radio spectrum resource. Section VI, then, provides a forecast of technological and service trends beyond the medium term trends dealt with in Section IV. It is intended to provide readers with a longer term, albeit more speculative, view of trends in wireless industry. Finally, Section VII provides a summary of the report and offers some preliminary conclusions regarding the impact on the accessibility of the wireless systems and services for people with disabilities.

Regulatory Framework and Industry Structure

Under powers delegated to it by the Congress, the Federal Communications Commission (FCC) has the exclusive responsibility and authority to allocate, allot, assign and otherwise manage the use of the radio spectrum resource by private individuals, commercial entities and state and local governments in the United States. The President of the United States has corresponding responsiblity and authority over the Federal government's own use of the resource. The President, in turn, has delegated certain of these responsibilities for Federal government use of the spectrum to the National Telecommunications and Information Administration (NTIA), a unit of the U.S. Department of Commerce.

In addition to its responsibilities and authority relating to the radio spectrum resource, the FCC also has broad authority to regulate interstate and international common carrier communications facilities and services (e.g., ordinary telephone services provided by local exchange and interexchange carriers). The FCC's authority over telecommunications facilities and services is shared with state public utility commissions (PUCs) who have regulatory jurisdiction over intrastate common carrier facilities and services. Thus, for example, the state PUC has jurisdiction over the price charged for an ordinary telephone call from one city to another city within the same state, while the FCC has jurisdiction over an ordinary telephone call from one city to a city in a different state. In the case of the FCC, its governing legislation is the Communications Act of 1934, as amended.

As a result of a 1993 amendment to the Communications Act of 1934, two categories of of mobile (wireless) telecommunications services were created: Commercial Mobile Radio Service (CMRS) and Private Mobile Radio Service (PMRS). CMRS includes all mobile radio services that are provided for a profit, are interconnected with the public switched telephone network, and are available to the public at large. PMRS includes any mobile radio service that is not CMRS or its functional equivalent. Thus, for example, a cellular mobile radio service is classified as CMRS while a two-way mobile radio system owned by a taxicab company and used to communicate with its own fleet of taxis would be classified as PMRS.

Based upon the description provided earlier, it might appear that local calls using a CMRS provider (e.g., a cellular telephone company) would be under the jurisdiction of the state PUC. However, in the same legislation that created the CMRS/PMRS regulatory framework, the Congress explicitly preempted state and local rate and entry regulation. Under this regulatory scheme, CMRS providers are subject to various regulations as common carriers by the FCC rather than by state PUCs. Generally speaking, regulation of PMRS is limited to engineering/technical factors (e.g., maximum allowed transmitter powers). Specifically, they are not subject to economic regulation (i.e., control over their prices and terms and conditions of their services) as are common carriers.

Thus, FCC regulation is particularly important in terms of wireless telecommunications because of (a) the agency's control over the radio spectrum resource and the equipment that uses it and (b) the agency's jurisdiction over commercial providers of mobile (wireless) telecommunications services. In terms of the community of persons with disabilities, the FCC is particularly important because of the agency's broad authority to establish technical standards — standards that impact on the accessibility and usability of not only wireless telecommunications facilities and services, but wireline facilities and services as well. Even more directly, the FCC's role is crucial because of its responsibilities to carry out those portions of the Communications Act of 1934, as amended, that deal specifically with access to telecommunications by persons with disabilities.

THE FUNDAMENTALS OF WIRELESS COMMUNICATIONS

Principles of Wireless Communications

Radio, or the use of radiated electromagnetic waves, is the only practical way of communicating with people or vehicles that move around on land, on the sea, in the air, or in outer space. It is the use of electromagnetic waves that permits the transmission and reception of information over a distance without the use of wires. The distance covered may range from only a few feet in the case of a cordless telephone to millions of miles in the case of a space probe.

In principle, radio communications is a relatively straightforward process. At the transmitting or sending end, the information to be sent (e.g., a voice signal) is imposed on a locally generated radio frequency (RF) signal called a carrier. The process of imposing the information signal on the carrier is called modulation. This carrier signal, along with the information signal imposed on it, is then radiated by an antenna. The frequency of an electromagnetic or radio wave is simply its oscillation rate measured in cycles-per-second or Hertz. The range of radio frequencies useful for practical communications starts at a few thousand Hertz (Hz) and goes up to a few hundred billion Hertz.

At the receiving end, the signal is picked up by another antenna and fed into a receiver where the desired carrier with the imposed information signal is selected from among all of the other signals impinging on the antenna. The information signal (e.g., voice) is then extracted from the carrier in a process referred to as demodulation. Thus modulation of the carrier wave occurs at the transmitter (the emitter of the radiation) and demodulation occurs at the receiver. These same basic steps or processes can be identified in radio systems ranging from the cheapest cordless telephone with a very low power transmitter and simple antenna to a high power transmitter carrying multiple information signals and utilizing complex, directive antennas.

A pure, unmodulated radio carrier conveys no information and occupies only an infinitesimal amount of the spectrum. Modulation of the radio signal inevitably causes a spreading of the radio wave in frequency. Thus a radio signal conveying information occupies a range of frequencies called a channel. In general, the more information that is sent per unit of time, the wider the channel must be.

As the radio wave expands in surface area after leaving the antenna, it grows weaker and weaker. At the receiver, the signal must still be strong enough to overcome any local radio noise or interference; otherwise the transmission will not be successful. In outer space, where there are no intervening hills or mountains, natural foliage, or man-made objects such as buildings with which to contend, the weakening of the signal with distance from the transmitting antenna can be predicted with great precision. In terrestrial radio systems, the environment for transmission is much more complex. It is even worse in mobile systems — where one or both of the terminals (transmitters and receivers) can move about — due to an environment that changes dynamically from moment to moment.

One major effect that appears in the terrestrial environment is multipath. Multipath is produced when the radio wave not only travels directly from the transmitting antenna to the receiving antenna, but is also reflected off of other physical objects such as buildings or mountains. At some locations, the signals traveling by different paths may add up to make the signal stronger, while at other locations, just a short distance away, the signals can cancel one another, causing the signal to fade. This effect is referred to as multipath fading. In addition, a large building or mountain between the transmitter and receiver may block the signal entirely, producing another type of fading.

Thus, in a terrestrial mobile environment, the communications engineer must not only take into account the natural weakening of the signal with distance (the so-called free space loss), but also the rapid changes in signal strength caused by multipath fading, the fading caused by shadowing, as well as the additional weakening of the signal produced when customers use portable units inside buildings or vehicles.

In doing so, the communications engineer uses complex computer models and field measurements to determine design parameters (e.g., transmitter power and antenna heights) to ensure that service is adequate over the desired coverage area.

As stated above, in addition to extracting the information from the radio wave through demodulation, it is also a principle function of a receiver to accept only the information in the chosen channel and reject other information being sent simultaneously in other (e.g., adjacent) channels. The measure of the receiver's ability to reject interfering signals on other channels is referred to as its selectivity. Hence, two or more radio systems can use the radio spectrum in the same area at the same time as long as (a) they are separated sufficiently in terms of frequency — i.e., so that their channels do not overlap, and (b) the receivers involved have sufficient selectivity to reject the signals on adjacent channels.

If two radio systems do occupy the same channel, they must either time share the channel in some way or be separated enough in distance to not cause interference to one another at the desired reception points. In other words, the receiver must be close enough to the desired transmitter location, and far enough from the undesired or interfering transmitter, to ensure the strength of the desired signal relative to the strength of the undesired signal is great enough to provide the needed quality. Generally speaking, because of the drop in signal strength with distance, the further the receiver is from the desired transmitter, the further away the undesired transmitter must be to prevent harmful interference.

In a traditional, two-way radio system used by taxicab companies, for example, where the desired radius of coverage around the base station transmitter is, say, 20 miles, the interfering transmitter must be something like 70 miles away. This is often referred to as the frequency reuse distance. As will be described in more detail in later sections, more intense frequency reuse is extremely important in modern Personal Communications Service (PCS) systems as a way of increasing capacity. That is, by keeping the ranges and, hence the reuse distances, short, the same channel can be reused many times for different conversations in the same geographic area. In summary, at a very basic level, the radio spectrum resource can be shared by many simultaneous users by taking into account its frequency, space, and time dimensions. All of these dimensions are exploited heavily in modern wireless systems.

Types of Signals and More Details on Modulation

There are two basic types of signals — analog and digital. An analog signal is a signal that varies continuously between a maximum and minimum value. At a given instant, an analog signal can assume any one of an infinite number of values between the two extremes. Examples of analog signals include the human voice or other measurable values in the physical universe such as the temperature of a boiler. A digital signal, in contrast, does not take on a continuous set of values. Rather, at a given instant of time, it takes one of a limited set of values called a symbol. A sequence of such values or symbols can be used to represent a number or alphabetical characters. Examples of digital signals include the presence or absence of a current pulse on a wire or a light pulse on a fiber optic cable. In this example, the pulses can be interpreted as binary digits or bits, and particular sequences of bits can be uniquely defined to correspond to numbers or alphanumeric characters. A communications system can be either analog or digital (or a combination of the two); that is, the information can be transmitted in either the analog or digital form within the network itself.

As described before, the carrier signal in a radio system is characterized by its frequency measured in Hertz. In addition to its frequency, the carrier is also characterized by the amplitude or strength of the wave and by its phase. Modulation of the carrier wave is accomplished by varying any or all of these characteristics in a known relationship to the information signal. For example, the amplitude of an analog information signal can be used to vary the amplitude of the carrier wave in a process known as amplitude modulation. Or the amplitude of the analog information signal can be used to vary the frequency of the carrier wave in a process known as frequency modulation. At the receiver, these amplitude or frequency variations in the carrier wave are used to extract the information signal in the demodulation process. These are examples of an analog communications system. In ordinary amplitude modulation, the channel width must be twice the highest frequency present in the information signal. In frequency modulation, the channel width is typically several times the highest frequency present in the information signal.

In the land mobile radio field, which is the focus of this report, the predominant modulation technique has been frequency modulation. At first glance, it might appear that frequency modulation makes inefficient use of the radio spectrum resource since the channel width required is much greater than for amplitude modulation. However, the actual situation is much more complex because, as a general proposition, signals that are spread over wider channels are more resistant to noise and interference. Without going into a lot of technical details, suffice it to state here that the wider the signal being transmitted relative to the width of the information, the greater the ability of the system to suppress noise and interference (including multipath). Thus, there is a tradeoff between transmitted bandwidth and noise and interference resistance. This improvement in performance is particularly useful in (a) the hostile radio signal environment described earlier and (b) rejecting interference from distant transmitters.

Digital signals can also be transmitted over radio systems by varying any of the three parameters described — frequency, amplitude or phase. The earliest form of digital modulation, Morse Code, simply turned the transmitter on and off to form dots and dashes that could be interpreted by human operators as symbols (e.g., letters). The on and off pulses or bits that comprise a modern digital signal could be sent the same way, i.e., by turning the transmitter on to signify a one and off to signify a zero. But because of the ever-present fading on radio paths, the receiver would not be able to reliably determine whether a zero had been sent or the signal was simply in a fade. Thus the more reliable way is to transmit one frequency to signify a one and another frequency to signify a zero. This is referred to as Frequency Shift Keying (FSK). Modern digital systems use combinations of frequency, amplitude and phase modulation to increase the number of bits that can be transmitted in a given channel.

As will be described in more detail later, the trend in wireless systems (just as in wireline networks) is toward digital systems and the use of advanced forms of digital modulation. Digital systems have a number of important advantages including the fact that digital signals are more immune to noise and, unlike analog systems, even when noise has been picked up, any resulting errors in the digital bit stream can be detected and corrected. Moreover, digital signals can be easily manipulated or processed in useful ways using modern computer techniques. While it is easy to envision how digital information signals are sent over digital communications systems, the method of sending analog signals (like voice) over a digital communications system and reproducing them at the other end is not as obvious.

In a digital system, the analog signal is digitized in an analog-to-digital converter to make it compatible with digital transmission. That is, the analog signal is converted into a sequence of bits that accurately describe the analog signal. More specifically, (a) the analog signal is sampled at sufficiently close intervals to accurately reproduce the signal's shape, (b) the amplitudes of the samples (the amplitude of the analog signal at particular instants of time) are quantized or given an approximate value according to the range within which the amplitude falls, and (c) these amplitude values are then encoded as a sequence of bits representing the corresponding binary number. At the receiving end, the analog signal is reconstructed from the sequence of bits that describe the amplitude of the signal at each instant of time. Thus, in a digital cellular system, for example, a voice signal is first converted into a digital signal and is then carried over digital transmission facilities that employ one of the advanced forms of digital modulation of the carrier wave described above.

The technique for converting an analog signal to a digital signal as just described is known as waveform coding. One popular form of waveform coding, called Pulse Code Modulation converts speech into a digital bit stream operating at 64,000 bits per second (bps) or 64 kbps. Another form of waveform coding is known as Adaptive Differential Pulse Code Modulation (ADPCM) and it operates at 32 kbps. It is possible to reduce the number of bits per second it takes to describe a voice signal by taking advantage of the known characteristics of the human voice. These techniques are known as voice coding and the devices employed to implement the techniques are called vocoders. It is beyond the scope of this report to describe the different techniques used for voice coding; suffice it to state that these techniques all use computer processing power to remove redundancy from speech so that (a) fewer bits per second have to be transmitted to convey a voice signal and (b) the available bandwidth can be used more efficiently. Some of these techniques allow voice to be sent at rates as low as 4 kbps (and even lower), compared with the 64 kbps or 32 kbps associated with waveform coding.

At very low rates, the quality deteriorates and the reproduced voice signal takes on a computer generated-like sound. In addition, the compressing and decompressing of the signal takes time, even with fairly powerful processors. At high levels of compression, the resulting delays can be annoying to end users. Thus, there is a tradeoff between the bit rate, the quality of the voice reproduction (including delay), and the amount of computer processing power employed within the transmitter and receiver. The latter not only has implications for cost, but also for battery life as well, since more processing power translates into increased battery drain. This is a particularly important consideration in portable units.

Most operators of commercial wireless telephone systems have a strong incentive to employ digital voice compression because the lower bit rates translate into (a) more conversations in a given amount of spectrum — i.e., more efficient use of the radio spectrum and greater capacity, and (b) more conversations per piece of radio equipment and/or radio site — i.e., greater economies of scale. Moreover, the FCC generally encourages the increased spectrum efficiency that results from voice compression. It is important to stress that the use of voice compression can cause problems for non-voice signals such as those emitted by fax machines, computer modems, and, especially important in the context of this report, text telephones/TTYs. This is because, as described above, vocoders depend critically upon the signal having the characteristics of the human voice. While the range of audio frequencies is the same, anyone who has listened to a fax machine, modem, or TTY on a telephone line knows that it does not sound like the human voice. The use of vocoders in wireless telecommunications and the implications of that use for the deaf community will be discussed in more detail in later sections.

Licensed Bands Available and Their Technical Characteristics

Domestically, the FCC has set aside certain bands or ranges of radio frequencies for land mobile radio use. These bands include Low Band in the 40 MHz region of the spectrum, High Band in the 150 MHz region, a band near 220 MHz, the UHF band in the 450 MHz region, a band near 800/900 MHz, and, most recently, a band near 1.9 GHz. With the exception of the 220 MHz band, which was recently set aside for land mobile radio use, the FCC has historically opened up bands higher up in the spectrum as the lower bands have become more congested. Thus, following World War II, most wireless mobile activity was centered in Low Band. However, in response to rapid growth in the land mobile radio service, the FCC, over the intervening years, has steadily increased the amount of spectrum available through successive reallocations of the resource in the higher frequency ranges.

Moving higher in frequency to avoid congestion has advantages and disadvantages. Generally speaking, the radio frequency (RF) devices employed within the system get more costly the higher the frequency and, in terms of propagation effects, the higher frequencies are subject to more blocking or shadowing by buildings or hills. However, the higher frequencies tend to penetrate buildings more readily and the antennas involved are physically smaller — both important attributes for systems that seek to serve small portable units carried on one's person. At some risk of over-generalizing, it can be said that (a) the lower frequency bands are best for economically covering wide areas in suburban and rural areas where frequency reuse is not as important and (b) the higher frequency bands are best for covering urban areas where building penetration and high levels of frequency reuse are desired.

Multiple Access and Duplexing Techniques

If spectrum were unlimited and the radio equipment used in the infrastructure were free, everyone could have their own wireless channel within one of the bands set aside by the FCC for land mobile radio use. But spectrum is not unlimited and the backbone equipment is not free. Thus it is imperative that the spectrum and, often, the backbone equipment, be shared among users. In short, users in a given area must contend for a limited number of channels.

There are different ways of dividing up the spectrum and providing users access to it in an organized way. The simplest and most straightforward method is known as frequency division multiple access (FDMA). With FDMA, the available spectrum is divided into non-overlapping slots in the frequency dimension or domain. These frequency slots or channels are then put into a pool and assigned to users on either a manual or automated basis for the duration of their particular call. For example, a 150 kHz block of spectrum could be divided into 6 channels or frequency slots each 25 kHz wide. Such an arrangement would allow six simultaneous conversations to take place, each with their own carrier within their own frequency slot. In the example, this would mean that each user would be continuously accessing one-sixth of the available spectrum during the duration of the conversation. FDMA is perhaps the most familiar way of dividing up spectrum, and it has traditionally been associated with analog systems.

With TDMA, the available spectrum is divided into non-overlapping time slots in the time dimension or domain. These time slots or channels are then put into a pool and assigned to users for the duration of their particular call. To continue the example given above, in a TDMA system the 150 kHz of spectrum would be divided into recurring groups (frames) of six time slots, and each time slot would carry a sequence of bits representing a portion of one of six simultaneous conversations. The six conversations each take turns using the available capacity. In other words, each user would be accessing all of the available spectrum but only for one-sixth of the available time. Rather than each signal having a particular frequency slot as in FDMA, in TDMA each conversation occupies a particular time slot in a sequential fashion. The frames are repeated fast enough that there is no interruption or delay in the conversation as seen by the end user.

Note that, theoretically at least, there is no difference in capacity between FDMA and TDMA as seen by the end user. Namely, you get access to one-sixth of the capacity all of the time or all of the capacity one-sixth of the time to continue the example. Note further that, in the practical world, digital systems are typically a combination of FDMA and TDMA. In other words, the systems are designed so that the capacity is divided into both the frequency and time dimensions whereby a user contends for a particular channel and then a time slot within that channel.

A third access method is known as Code Division Multiple Access (CDMA). CDMA is both a modulation and an access technique that is based upon the spread-spectrum concept. A spread-spectrum system is one in which the bandwidth occupied by the signal is much wider than the bandwidth of the information signal being transmitted. For example, a voice conversation with a bandwidth of just 3 kHz or so would be spread over 1 MHz or more of spectrum.

In spread spectrum systems, multiple conversations (up to some maximum) simultaneously share the available spectrum in both the time and frequency dimensions. Hence, in a CDMA system, the available spectrum is not channelized in frequency or time as in FDMA and TDMA systems, respectively. Instead, the individual conversations are distinguished through coding; that is, at the transmitter, each conversation channel is processed with a unique spreading code that is used to distribute the signal over the available bandwidth. The receiver uses the unique code to accept the energy associated with a particular code. The other signals present are each identified by a different code and simply produce background noise. In this way, many conversations can be carried simultaneously within the same block of spectrum.

Before going on to discuss the types of services in the wireless field, one further technical topic must be addressed, and that is "duplexing." In many, if not most, communication systems, it is desirable to be able to communicate in both directions at the same time. This system characteristic, which is known as full-duplex operation, is desirable because it lets one party in a voice conversation interrupt the other with a question or one device to immediately request a retransmission of a block of information received in error during a data communications session. There are two basic ways of providing for full-duplex operation in a radio system. By far the most common is to assign two different frequency slots per conversation — one for transmitting and one for receiving. By separating the slots sufficiently in frequency, filters (say in the portable radio) can be used to prevent the transmitted information from interfering with the simultaneously received information. Thus, in many land mobile radio bands, a channel actually consists of two frequency slots — one for each direction of transmission in a full-duplex conversation. This arrangement is called Frequency Division Duplexing (FDD).

Another much less common means of achieving full-duplex operation in the digital world is through what is called time division duplexing (TDD). In TDD, a single (unpaired) channel is used with each end taking turns transmitting. Each end sends a burst of information (consisting of bits representing a few samples of the voice signal, for example) and then receives a burst from the other end. As in the case of the TDMA access technique, this process is repeated rapidly enough that the end user does not perceive any gaps or delays in what is heard. To the end user it appears as a true full-duplex connection.

Types of Services

Although there are many sub-markets and niches, the land mobile market can be divided into four traditional segments serving four different applications. These four segments or applications are (a) one-way paging or messaging, (b) two-way dispatch, (c) two-way mobile/portable telephone, and (d) two-way data or messaging. Understanding these different applications is important from a technological perspective because they all have different requirements. This means, among other things, that a network or system optimized for one application may not be optimal for another. Thus, one can observe in the marketplace standalone systems optimized for dispatch service, for paging, for interconnected mobile telephone service, and for two-way data/ messaging. One can also observe systems that attempt to capture economies of scope by offering combinations of these services on a common infrastructure. In the following few paragraphs, each application will be briefly described.

One-way paging or messaging uses a radio signal to merely alert or to instruct the user to do something. The user (an office equipment repair person or a doctor, for example) carries a very small device to receive the one-way messages. Often this device is referred to as a "pager" or a "beeper." There are four types of paging services — tone-only, tone-voice, numeric, and alphanumeric. In the tone-only system, the receiver simply emits a tone which alerts the user to take some predetermined action such as calling their office or answering service. In the tone-voice system, the tone is followed by a short voice message entered by the person placing the page. In the numeric system, a short numeric message is sent and displayed on a small screen on the receiver. A typical numeric message might be the telephone number that the user is supposed to call. An alphanumeric system is similar except that it allows a more complex text message to be delivered. Paging, like cellular mobile radio systems, has exhibited rapid growth.

Two-way dispatch is another basic land mobile radio service. It involves communications between and among a dispatcher and units (mobiles and/or portables) in the field. It is typically a "command and control" system where a high degree of coordination among the units is required. Such services are used heavily by the public safety community and by businesses like tow truck and taxicab companies that must dispatch units operated away from the principal place of business. There is typically a requirement for the dispatcher to be able to reach multiple units simultaneously in what is referred to as group or fleet calling (i.e., one-to-many communications). The messages are typically of short duration (tens of seconds) and efficiency and other considerations dictate rapid call setup. Push-to-talk and release-to-listen (PTT/ RTL) and half-duplex operation are common. In its pure form, dispatch communications does not involve interconnection with the Public Switched Telephone Network (PSTN) and, in many applications, such interconnection is neither needed nor desired.

Two-way mobile telephone is another basic land mobile radio service. It allows the user to place and receive ordinary telephone calls (i.e., one-to-one communications) and, obviously, provision must be made for interconnection with the PSTN. The messages are typically of much longer duration (compared to dispatch calls) and users typically demand full-duplex operation. Because the call itself is of longer duration, call set-up delay is less critical. This service need not be described in detail, since the basic notion is to duplicate the operation of the ordinary telephone network, but with wireless telephones or handsets. The fourth and final basic land mobile radio service is two-way data messaging. This service is of more recent origin. It facilitates various forms of data communications such as computer aided dispatch, electronic messaging/ mail, telemetry, and computer-to-computer communications on a wireless basis. The data traffic on such networks is typically "bursty" in nature, and errors cannot be tolerated in many critical applications. On the other hand, unlike with voice applications, the systems are typically tolerant of transmission delays of up to several seconds. These wireless data communications services are used, for example, by package delivery services to track packages and to schedule pickups and deliveries.

Customers typically have differing requirements for the four services. Some users may only need one-way paging or mobile telephone service, some may need dispatch and mobile telephone, while others may have a need for all four. Systems to provide these services typically started out on a separate, standalone basis and systems are still evolving. At the same time, as mentioned earlier, systems are also evolving that try to offer an integrated set of services on a common infrastructure or platform.

It should be pointed out that these services can be (and are) provided on both a private and third-party (e.g., common carrier or other commercial) basis. For example, a user can purchase and operate a radio system and provide dispatch communications to its own fleet or purchase the service from a Specialized Mobile Radio operator who provides the services on a commercial, for-hire basis. As noted in Section I.D., third parties who offer interconnected services on a commercial, for-hire basis are categorized as Commercial Mobile Radio Service (CMRS) providers.

TRADITIONAL LAND MOBILE RADIO SYSTEMS

The purpose of this section is to describe the systems and technologies traditionally used in the provision of the different types of services described in Section II. It also deals with the limitations of these systems in order to provide the reader with a better understanding of the impetus for the development and deployment of modern systems described in Section IV which follows. The distinction between "traditional" and "modern" systems is an arbitrary one, but it does provide a means for understanding the latest developments in the field. Because the wide-spread deployment of two-way mobile data systems is a fairly recent phenomenon, they will only be discussed in Section IV.

Paging

A paging system with dial-in capability consists of a terminal and a radio distribution network. The terminal, which contains computer logic, answers the incoming line and, based upon the telephone number dialed, matches that number to the corresponding pager address stored in memory. If the service is other than tone-only, the terminal prompts the caller for some additional information (e.g., a voice message if it is a tone-voice service, a telephone number if it is a numeric display service, or free form text for an alphanumeric display service).

In its simplest form, the radio distribution network may consist of a single transmitter and associated antenna mounted on a tower, building, or mountain (or a combination of these). In this simple configuration, the terminal merely puts the message into the correct format or protocol for transmission and conveys it to the transmitter where it is broadcast over the coverage area. The pager address may be broadcast as a sequence of bits (ones and zeroes) using digital modulation. The address is followed by the message itself if it is other than the tone-only service. The paging receivers (pagers) in the coverage area listen for their address and, if they "hear" their unique address, they are activated and the trailing message is delivered. As explained in more detail below, a modern paging system may employ a network of multiple transmitter locations allowing an extension of the geographic service area.

Single Frequency Dispatch

The simplest type of two-way dispatch system uses a single frequency slot (i.e., an unpaired channel) for both transmitting and receiving. The parties at each end of the conversation (e.g., a dispatcher and a mobile unit in the field) take turns talking. This is sometimes referred to as the push-to-talk/release-to-listen (PTT/RTL) mode. The base station consists of a simple transmitter/receiver combination (a transceiver), an antenna mounted on the building or tower, a transmission or feedline (normally coaxial cable) for connecting the transceiver to the antenna, plus, perhaps, some accessories such as a dispatch console. These simple dispatch systems typically operate in the analog mode and employ Frequency Modulation (FM).

These conventional, single frequency dispatch systems are still used extensively at both low-band and high-band. Two fundamental disadvantages of such systems are (a) that they mix higher power base stations and lower power mobile units in the same frequency slot which exacerbates interference problems when the channel is shared among multiple systems, as is often the case, and (b) that they do not permit the use of repeaters (as described below), nor do they allow for full-duplex operation. Other disadvantages of these systems will be described in Section III.D.

Two-Frequency Dispatch/Community Repeater

As explained earlier, over the years, as low-band and high-band became more congested, the FCC regularly allocated (or reallocated) additional spectrum at higher frequencies for use by the land mobile radio services. This meant that a simple, single frequency dispatch system with the antenna on the user's premises often did not provide adequate range, especially for communicating with much lower power portable radios that were gaining in popularity. This led the FCC to organize the bands to permit the assignment of paired frequency slots for each channel. Thus, other than for one-way paging, paired frequency assignments are available for dispatch systems in the UHF and 800/900 MHz radio bands.

This arrangement allows the higher power stations to transmit on one frequency of the pair, and the lower power mobile units on the other, thus eliminating the power disparity and reducing interference. It also allows the deployment of mobile relay or community repeater systems. In community repeater systems, a relatively high power transceiver and associated antenna are placed on a tall tower, building or mountain top that offers good coverage over a wide area. The mobile units transmit on one frequency which is picked up by the receiver in the unit on the mountain/building top and then simply retransmitted or repeated on the other frequency of the pair. This special type of transceiver with the receiver and transmitter hooked "back-to-back" is called a repeater. With such a system, the dispatcher can communicate with his/her mobile units via the repeater using a low power transceiver located on his/her premises. The latter is known as a control station. Thus, the repeater relays all of the low power signals, enabling both control stations and mobile units to communicate with one another over a wide area.

These systems typically operate in the PTT/RTL mode, although full-duplex operation is possible when communicating to and from the repeater (e.g., when telephone interconnection is provided at the repeater site). In addition to wide coverage, repeater systems offer other advantages. For example, because a repeater simply retransmits what it "hears," it is very easy for several customers with different fleets and their own control stations to share the use of a single repeater. This means that individual users with a small number of mobile units do not have to construct their own dedicated, high power/high antenna site facilities. The systems also allow the use of low power mobile and portable units with corresponding reductions in cost. Like the single frequency dispatch systems, these two-frequency, repeater systems operate in the analog mode and employ FM modulation.

DISADVANTAGES OF SINGLE FREQUENCY AND PAIRED FREQUENCY REPEATER SYSTEMS

The two types of two-way dispatch systems discussed thus far exhibit several basic limitations. These limitations are briefly described in the paragraphs which follow.

Undisciplined Access to a Shared Channel

As explained earlier, there are not enough channels available to allow each user to have a dedicated channel of his or her own. Thus, except in the case of the very largest users, the channels must be shared among a number of such users. For example, a particular UHF channel might be shared among five companies, each with a control station and five to ten, or even more, mobile/portable units each. Because each individual user may not know when another user is going to transmit (i.e., push the PTT button), users will occasionally interfere with one another. As congestion grows, the situation gets worse and there may be several users each waiting to transmit and each unaware of the others. Moreover, some less polite users may get impatient, and transmit even if someone else is on the channel. Thus the inevitable consequence is delays, interference, and lost messages, especially when there are unaffiliated users on the channel. In short, access to the channel is not disciplined.

Limited Addressing Capabilities and Lack of Privacy

Both of the described systems operate as giant party lines in which each user is generally able to hear the conversations of other users. Some systems are equipped with relatively simple devices that allow messages to be directed to a particular mobile or group of mobile units. However, these addressing systems are rather rudimentary and, while they help reduce the amount of chatter to which a particular user has to listen, they do nothing to protect the user against casual eavesdropping by other people sharing the channel or by individuals (including competitors) employing simple scanning receivers. Moreover, they typically do not have enough unique addresses to allow the creation of large, networked systems.

Severe Channel Congestion in Some Areas/Services

Even with the creation of new land mobile radio bands, channels are often very congested during peak periods, especially in major urban areas like New York and Los Angeles. This often produces excessive delays in accessing the channel.

Inefficient Use of the Spectrum Resource

Despite the heavy loading, at any instant of time, there may be some channels in the area that are unused or lightly used because, for example, the peak usage of the different channels may not coincide. However, with the simple systems just described, access is limited to only one channel or only a handle of manually selected channels. It is clear that it would be more efficient for many channels to be put into a common pool and then drawn from the pool to carry conversations on an as-needed basis. In other words, in technical terms, the channels are not efficiently used because they are not "trunked."

Moreover, because they use high power and high antenna sites, a single conversation — say between a control station and a mobile unit a few miles apart — precludes the use of a channel over a very wide geographic area. For example, a system providing coverage over a 20 mile radius may preclude the assignment of the same channel for some 70 miles. In short, the systems are not spectrally efficient because the channels are not reused intensively in a given area. Finally, the channels are not efficiently used because a single voice conversation with a nominal bandwidth of 3 kHz occupies a 25 kHz frequency slot in the radio spectrum.

MULTICHANNEL TRUNKED RADIO SYSTEMS

Multichannel trunked systems, such as those operated by very large private organizations on a private basis and by third-party providers on a commercial basis, are designed to provide more disciplined access to the channels and to allow for the message-by-message sharing of a pool of channels. They operate on an FDMA basis. In a modern multichannel trunked mobile radio system, this is accomplished through the use of computer logic which assigns channels from a pool and recovers them at the end of a transmission or message. Thus a modern trunked mobile radio system consists of a collection of repeaters, each operating on one of a pool of multiple channels (typically from five to 20), and under computer logic control.

The FCC did not require the standardization of trunked radio systems and a number of proprietary systems have emerged. In one popular system, one of the available channels in the pool is set aside as a digital signaling or control channel. All of the end-user mobile units and control stations monitor the control channel and make requests and receive instructions on it. When a user indicates that he or she wants to send a message by pushing the PTT button, the mobile unit or control station sends out a burst of digital information on the control channel which identifies the individual mobile unit or fleet of mobile units with which the user wants to communicate. The computer logic at the repeater site finds an idle channel in the pool and sends back a burst of digital information telling the individual mobile unit or fleet of mobile units to all move to the selected idle channel. Once the control station/mobile units have arrived on the idle channel, the user who originated the transmission can begin to talk. In a modern system, this whole process takes less than one-half second.

If all of the channels are busy, the call requests are placed in a queue and handled on a "first-in, first-out" basis. At the end of the conversation, the channel is returned to the pool and mobile units and control stations in the fleet go back to monitoring the control channel again. All of this is done automatically and all channels are available to all users. Note that the conversation channels are only used for the duration of the call.

Trunked mobile radio systems can be used for placing and receiving ordinary telephone calls by interconnecting the transceivers at the repeater site with the Public Switched Telephone Network (PSTN). Thus, if interconnection is provided on the system, when a user wants to engage in a telephone call rather than a dispatch call, the conversation is appropriately routed to/from the PSTN. Trunked mobile radio systems overcome many of the disadvantages associated with the operation of individual repeaters such as Community Repeaters. First, they provide disciplined access to the channel which prevents people using the system from intentionally or unintentionally causing interference to other users. Second, because the audio output of the receiver portion of the mobile/portable radio is only activated when the unit's individual or group address is received, casual eavesdropping is eliminated. Moreover, because the assigned channel jumps around from conversation to conversation (or even from transmission to transmission), it is also more difficult for other people to monitor the conversations of a particular user. Third, the waiting time to access a channel is greatly reduced because, unlike in a single channel system, if any channel in the pool is idle, it can be immediately assigned for use in a conversation. Fourth, the radio spectrum is used much more efficiently because more mobile units can be accommodated per channel. For example, a modern trunked system providing dispatch service on a pool of say 20 channels can provide excellent service (i.e., short waiting times) with an average of well over 100 users per channel. As a rough estimate, a trunked system with a reasonable number of channels can provide about three times the capacity of untrunked channels for the same grade of service (i.e., average delay to access a channel).

The principal disadvantage of a trunked radio system of the type just described is that a single, point-to-point conversation between a control station and a mobile unit or between a mobile unit and the PSTN via the repeater site occupies a valuable radio channel over a very wide geographic area. In other words, little frequency reuse is employed, which means that the systems are spectrum inefficient for one-to-one calling. Another disadvantage in congested areas is that each end of the voice conversation occupies, typically, a 25 kHz channel because the systems use ordinary FM, analog modulation. In addition to these spectrum efficiency concerns, the wide coverage provided by the high power repeaters can cause difficulties in serving small, very low power portable units operating from within buildings and other difficult to serve locations. In other words, the imbalance in transmitter power may make it difficult for the portable unit to "talk-back" to the receiver at the repeater site. These disadvantages are being addressed in third generation trunked mobile radio systems that are described in Section V.

CELLULAR MOBILE RADIO SYSTEMS

Cellular mobile radio systems get their name from the notion of dividing a large geographic area (e.g., an entire metropolitan area) into a series of small, hexagonal shaped cells. The hexagonal cell was chosen as a conceptual tool because its shape roughly approximates the circular coverage of a base station and because, when they are fitted together, they completely cover the area. Unlike the large geographic areas associated with high power dispatch systems of the type described above, the area covered by an individual cell is much smaller — typically they have a radius of coverage from two to eight miles. Relatively low power base station transmitters and receivers (transceivers) with relatively low antennas are placed in each cell and connected by wirelines to the central switching computer called a Mobile Switching Center (MSC). The MTSO/MSC is, in turn, connected to the PSTN. The relatively low power/ low antenna heights "match" the coverage to the area of the cell.

The base stations communicate with the mobile and portable telephones in their respective cells. Because the coverage areas or cells are small, the same set of frequencies in one cell can be used in a distant cell within the same metropolitan area. Thus, a single conversation occupies a channel over only a small geographic area, and the same channel can be reused for another conversation in another cell within the same metropolitan area. Within the cells, access to the network is provided on an FDMA basis that is conceptually similar to methods used in the trunking systems described above. That is, the channels are trunked so that the same efficiency gains associated with trunking are obtained.

In order for cellular systems to function with acceptable intra-system interference, there has to be a way of controlling interference from one cell to another. Historically, and in most systems today, this was accomplished by dividing the available channels into separate blocks and assigning different blocks to different cells arranged in, for example, a seven-cell pattern. Within the pattern, adjacent cells use different blocks of channels, and this pattern is repeated over the geographic area in such a way that the reuse of channels occurs in cells that are separated sufficiently in distance to limit the interference to acceptable levels.

A key concept associated with cellular mobile radio systems is that a startup system covering a large metropolitan area can be built with large cells and, as demand develops, these large cells can progressively be divided into smaller cells. When the cells are large, a particular channel may only be used a few times (e.g., three times), while in a more mature system it might be used ten or more times. Thus, the system becomes progressively more spectrally efficient — i.e., more and more users are accommodated in the same amount of spectrum. To summarize, in a cellular system, increased spectrum efficiency comes from both trunking and extensive geographic reuse of channels. In the Community Repeater systems described earlier, reuse of a channel may be precluded over an entire metropolitan area, and they are spectrally inefficient in that sense. Moreover, in a cellular system, the small cells not only provide increased spectrum efficiency, they also provide better coverage to small, low power portable units since users are never far from a base station.

Because the cells are small, it is quite likely that, during the time span of an individual call, a user in a vehicle will move from the coverage area of one cell into the coverage area of another. Hence, besides setting up, maintaining, and tearing down calls, the MSC must manage call handoffs from cell-to-cell as the mobile units move throughout the coverage area. Note that while cellular systems are particularly efficient in terms of their use of spectrum for one-to-one calls, they are not particularly effective or efficient in handling one-to-many, fleet dispatch calls. This is true for two reasons. First, the setting up of a call on a cellular system is more complex and time consuming, hence making it inefficient for handling calls of very short duration as is typical of dispatch calls. Second, the individual mobile units comprising a fleet may be scattered over numerous cells requiring the use of multiple base stations and multiple channels to handle a simple one-to-many transmission.

Like the other types of land mobile radio systems described before, early cellular systems employed analog FM modulation and FDMA as the access technique. In the United States, the corresponding standard for the modulation, access technique, and other interfaces and protocols is known as the Advanced Mobile Phone System (AMPS). Modern wireless systems are being deployed using digital modulation and TDMA or CDMA as the access technique as explained in more detail in Section IV. The motivation for the shift to digital is primarily to extract more capacity from the existing spectrum space, to capture economies of scale, and to achieve increased functionality.

Cordless Telephones

One of the less publicized but important developments in wireless technology is the rapid growth in ordinary cordless telephones. Historically these telephones operated on an unlicensed basis and utilized very low transmitter powers. Hence their range is very limited. In the United States, the FCC set aside just ten, paired frequency channels for use by cordless telephones, and the quality of these analog systems often leaves something to be desired. Nevertheless, because they connect to the user's own telephone line, involve no "air-time" charges, are inexpensive, and offer mobility within or near a home or business, they have proven extremely popular. Some observers have pointed to this popularity as an indicator that the public has a high demand for wireless services.

MODERN WIRELESS SYSTEMS AND TRENDS

The purpose of this section is to describe new and evolving wireless telecommunications systems that are being developed and deployed (a) to take advantage of recent advances in hardware and software elements used in wireless systems, (b) to overcome the limitations associated with the traditional land mobile radio systems identified in Section III, and (c) to be responsive to changing end user demands and an increasingly competitive wireless telecommunications marketplace. As noted in the introduction, the focus of this section is on technological and service trends in the medium term

— i.e., developments that can accurately be foreseen based upon current developments in infrastructure and end user systems and equipment. Since many of these advancements are being enabled by advances in hardware and software elements used in wireless systems, the section begins with a review of such advancements.

ADVANCES IN ENABLING TECHNOLOGIES

On June 25, 1995, the Federal Communications Commission and the National Telecommunications and Information Administration established the Public Safety Wireless Advisory Committee (PSWAC) to provide advice on wireless communications requirements for public safety agencies through the year 2010. As part of the PSWAC process, a Technology Subcommittee was established to, among other things, identify emerging technologies that might serve to meet the needs of these agencies. The deliberations of PSWAC's Technical Subcommittee included presentations from nearly 20 organizations including manufacturers, service providers, organizations engaged in research and development, and users. In addition, many organizations with relevant knowledge and experience participated directly in the Technology Subcommittee's work. Although the work of the subcommittee related primarily to public safety communications, its efforts included an extensive review of advances in enabling technologies that are basic to all forms of wireless communications — not just public safety applications.

In its final report, the Technology Subcommittee identified and described advances in nine enabling technologies: digital integrated circuits, RF generation devices, source coding, modulation, multiple access techniques, error correction coding, constraints on the use of various bands, backbone system elements, and performance modeling and verification. The following discussion summarizes and builds upon the work of the Technology Subcommittee.

Digital Integrated Circuits

In its report, the PSWAC Technology Subcommittee observed that the fundamental technology thrust through the year 2010 will continue to be, as it has been in the recent past, that of semiconductor technology. This fundamental technology thrust is a "two-edged sword." On the one hand, it increases the need for various computer-based services and, hence, increases the demand for radio spectrum to accommodate them. On the other hand, increased semiconductor capabilities make possible improved information compression techniques and more efficient modulation techniques that facilitate more efficient use of the radio spectrum resource and, hence, at least partially offset the need for additional spectrum. These same advances in semiconductor technology also fuel increases in the "intelligence" or computer processing power residing in the wireless network itself and in the end user equipment used to access the network. The Technology Subcommittee quantified past improvements in memory devices, microprocessors, and computer systems and projected them into the future. The impact of these trends in semiconductor technology is analyzed in more detail in later sections.

RF Generation Devices

The PSWAC Technology Subcommittee included batteries, oscillators, and antennas under this heading. In terms of battery technology, the Technology Subcommittee observed that batteries required to operate portable communications equipment are usually heavy, provide limited operating time, and can be expensive. It went on to note, however, that a number of developments in battery technology are alleviating this situation. These involve new technologies such as nickel-metal-hydride and lithium-ion batteries as well as zinc-air batteries that draw oxygen from the atmosphere to extend their life. It also noted that improvements were being made in wireless systems to conserve on the use of battery power. These include more efficient RF power amplifiers and more efficient "sleep" or standby modes.

In terms of oscillators, the Technology Subcommittee noted the ability to place more communications channels with a given amount of spectrum depends upon both the transmission bandwidth and the stability of the oscillator. In other words, if the oscillator drifts, for example, to changes in temperature, then more guard space must be provided between channels to prevent spillover or interference from one channel into the next. The Technology Subcommittee quantified past improvements in oscillator stability and the impact on spectrum efficiency. These improvements in oscillator stability, coupled with the advances in semiconductor technology described above, permit the design of radio equipment with the ability to operate on a frequency agile fashion on literally hundreds of different channels.

In terms of antennas, the Technology Subcommittee, among other things, noted the development of "smart antennas." Essentially, these antennas utilize microprocessor technology to electronically (rather than physically) steer the radio beam at the transmitter and/or receiver site. Such antennas can be used to reduce interference and improve performance by, for example, allowing the base station antenna to track a low power portable unit. The Technology Subcommittee noted that, while such techniques have been used in military systems for some time, they have not been widely used in commercial systems because of cost considerations. They noted, however, that this was likely to change with attendant improvements in semiconductor technology.

Source Coding

In Section II, waveform coding and voice coding were both described in the context of voice communications. The term source coding is a general term for techniques that take into account the known characteristics of the information source and receptor (e.g., the human ear or eye) to reduce the amount of information that must be sent over the communications channel. Typically these techniques work by removing redundant information or information that cannot be utilized because of limitations in the receptor. Voice coding, as described earlier, is one example of source coding, but analogous techniques can be used to compress other types of signals as well — including images (e.g., the transmission of facsimile messages or still pictures) and both slow scan and full motion video. Once again, it is beyond the scope of this report to delve into the details of these techniques but, as pointed out in the PSWAC Technology Subcommittee report, increasingly powerful digital signal processing integrated circuits will facilitate the introduction of more powerful and effective methods for reducing the amount of information that must be transmitted on a communications channel. The source content and compression capabilities of present day technology and expected gains in compression due to algorithmic advances and/or semiconductor technology gains are summarized in Appendix C of the PSWAC Technology Subcommitee report.

Modulation

Another technique for increasing the amount of information that can be transmitted in a given amount of bandwidth is to improve the modulation efficiency. As alluded to earlier, modern digital systems use various combinations of frequency, amplitude and phase modulation as well as other techniques to increase the number of bits per second that can be transmitted over a given channel. Implementation of these techniques is facilitated by the improved performance of digital signal processing integrated circuits. Modern wireless systems can achieve modulation efficiencies of over one bit per second per Hertz, even in the severe multipath/fading environment that is typical of mobile communications in an urban environment.

Multiple Access Techniques

Different channel access methods, including FDMA, TDMA, CDMA, and TDD, were discussed briefly in Section II. The PSWAC Technology Subcommittee report notes that these methods have specific strengths and weaknesses. It goes on to state that (a) FDMA is employed in narrowest-bandwidth, multi-licensed channel operation, (b) TDMA is employed in exclusive license use, moderate bandwidth applications and (c) CDMA is employed for widest-bandwidth applications in both single systems such as cellular mobile radio systems as well as uncoordinated and/or unlicensed applications (e.g., unlicensed, wireless local area networks). As described earlier, TDD is employed to achieve full-duplex operation in a single (unpaired) radio channel. While the Technical Subcommittee report briefly describes the advantages and disadvantages of each of these methods, it does not project any fundamental breakthroughs that would radically change or add to this basic set of multiple access techniques.

Error Correction Coding

As pointed out in the PSWAC Technology Subcommittee report, in a digital communications system, the objective is to maximize the ability of the receiver to successfully decode digitally encoded messages. In other words, the objective is to deliver without error the exact sequence of ones and zeros that was transmitted. The report states that:

A simplistic method to improve reliability is to send [the digital] messages more than once. This has the serious disadvantage of increasing transmission time by the number of times the message is repeated. More efficient methods uses [sic] error control techniques that add bits to the data stream in a precise fashion. The extra bits, however, are placed in a precise mathematically-prescribed pattern at the transmitter end such that complementary circuitry in the receiver can tell when an error has occurred, and determine what the correct bit value should be.

There are two types of error control techniques — simple error detection and forward error correction. Error detection is typically employed in data communications applications in concert with protocols that use a return channel to automatically request the retransmission of corrupted data. Forward error control provides the ability to detect and correct digital messages even in the presence of transmission errors. Forward error control is particularly useful for applications like voice where retransmission is not practical. While at first it may seem counter-intuitive that adding redundant bits would actually improve total performance, the increased throughput and/or decreased error rate associated with the use of sophisticated error control techniques more than compensates for the increase in transmission overhead. Such techniques are useful, for example, when improved performance cannot otherwise be achieved because of restrictions on maximum transmitted power. It should be obvious that, once again, these error control techniques are facilitated by the availability of increasingly powerful digital signal processing integrated circuits as described earlier.

Software Programmable Radios

The PSWAC Technology Subcommittee report contains a section entitled "Constraints on using various bands" but it is primarily devoted to software programmable radios. A software programmable radio is a radio in which functions or applications are configured under software control and in which the function or application, in whole or in part, is implemented in software resident in the radio. As pointed out in the Technology Subcommittee report, a software programmable radio requires the information or signal be presented in a digital format for processing. For example, even when receiving a digital radio signal, a traditional radio may perform many functions (e.g., filtering and demodulating the received signal) using analog components before the digital signal (i.e., the basic ones and zeros comprising the message) is actually extracted. In contrast, in a software programmable radio, these same functions may be accomplished using digital signal processing techniques. This requires, of course, that the received signal (including the carrier and the accompanying modulation) first be converted to the digital format.

Proponents claim a number of advantages for software programmable radios. One of the major advantages is that a number of different radios with different characteristics can reside or be hosted on one hardware platform, thus reducing the total amount of equipment required. For example, in principle, one could change the type of modulation used or the bandwidth of the channel employed by the radio as simply as one changes from a word processing program to a spread sheet program on a personal computer. In other words, such radios could facilitate multimode operation. In addition, the radio could be upgraded without changing the hardware, thereby increasing its performance at low cost. Similarly, software programmable radios could be used to facilitate the creation of multiband radios, and combining the multimode and multiband capabilities could facilitate interoperability among different radio systems. For example, it is conceivable that this high degree of functionality could be used to allow an end user to roam from one type of system — say one employing TDMA as an access technique — to another type of system — say one employing FDMA as an access technique. This would be accomplished by a change in software on a single hardware platform (radio) that allowed operations in both locations.

The Technology Subcommittee report notes that the military is placing an increasing emphasis on digital radio technology because of its potential for lower cost reconfigurability for multimode/interoperable communications. The report goes on to describe a military project for developing a software programmable radio whose goal it is to develop and open system architecture for radio service and demonstrate interoperability of multiple and simultaneous waveforms (signals) across a frequency range of 2-2,000 MHz. The report cautions that software radios are now much more expensive than hardware-based radios, but that "over time, the cost of software enabling technologies will decrease" and that "within a few hardware generations, software radios will sufficiently leverage the economics of advancements in microelectronics and provide seamless communications at a vest-pocket and palmtop level of affordability and miniaturization."

Backbone System Elements

In this section of its report, the Technology Subcommittee focused on a number of issues that are unique to public safety communications. But the basic notion that developments in backbone systems have implications for wireless telecommunications is an important one. Backbone systems are telecommunications systems, or, perhaps more accurately, subsystems that are used to carry signals between the base station sites and the control points and/or switch locations. The developments include lower cost, more reliable point-to-point microwave and fiber optic equipment for self-provision of backbone networks as well as an increasing number of commercial providers. For example, Competitive Access Providers (CAPs) are emerging in many metropolitan areas offering highly reliable, ground-based fiber optic connectivity. As described in more detail later in this report, there is also an increasing interest in more tightly integrating wireline and wireless networks wherein the wireless provider would not only share backbone systems with the wireline network, but switching, control, and "network intelligence" as well.

Performance Modeling and Verification

Radio systems engineers have an increasingly powerful set of computer-based tools for modeling, simulating, and measuring the performance of wireless telecommunications networks. While these tools are well hidden from the end user, they are extremely important to the proper design and operation of increasingly complex wireless systems. For example, improved software programs for predicting the coverage of radio systems and for modeling the interference between and among such systems can lead to significantly improved performance as ultimately seen by the end user.

ADVANCES IN WIRELESS SYSTEMS

Paging

There are three major technological advances associated with radio paging systems — the deployment of vast regional, national, and even international networks (including the use of simulcasting on a single frequency slot); the development and deployment of higher speed signaling techniques; and the development and deployment of two-way paging systems.

The deployment of paging systems covering large geographic areas has been made possible by the development of high-capacity terminals capable of controlling a large number of widely dispersed transmitters employing single or multiple frequencies and, often, offering a selection of coverage areas. Signals in nationwide and regional systems may be distributed from the central site containing the terminal to each market over satellite or terrestrial transmission facilities. In so-called "satellite" paging, signals are transmitted via satellite to very small aperture terminals (VSATs) located in each market served. Within a given market, a technique of simulcasting is typically used with multiple transmitters operating on a common frequency slot. Multiple transmitters are used both to extend the coverage area and to ensure good radio coverage, especially within buildings. With simulcasting, the transmitters and interconnection facilities are precisely controlled to prevent interference between the transmitters while still providing seamless coverage.

Another major advance in paging systems is the development and deployment of systems with protocols that operate at much higher speeds. Recent systems transmit one-way in a 25 kHz frequency slot at a speed of 6,400 bits per second (bps), a significant improvement over the prevalent standard of 2,400 bps. Moreover, the FCC recently allocated some new, wider paging channels (50 kHz) that allow transmission speeds as high as 112,000 bps. This not only provides for a vast increase in the number of pagers that can be supported per frequency slot or channel, it also allows for the provision of advanced services. For example, for many years, tone-voice services were discouraged in congested urban areas because the voice messages consumed a large amount of valuable airtime on the channel — time that could otherwise be used to transmit a large number of tone-only or numeric pages. The development of high-speed paging protocols and voice compression are now allowing providers to offer voice paging on an efficient basis, while eliminating the need for operators to transcribe voice messages into alphanumeric messages.

In addition to allocating wider channels for paging, the FCC recently made available (via auction) paging channels with a paired response channel. The services to be provided on these additional channels are sometimes referred to as Narrowband Personal Communications Service. The pairing of an in-bound frequency slot with an out-bound slot allows paging providers to offer two-way paging or messaging services. The response channel may be used to simply acknowledge the receipt of a page, to return one of a number of preprogrammed responses, or to provide a more symmetrical, two-way messaging service. In the latter case, the customer unit may no longer be a pager, of course, but rather a laptop computer or personal digital assistant (PDA).

Two-way Mobile Data

A fundamental advantage of mobile data systems is that they offer a much higher effective information transfer rate than voice systems. Human beings, dispatchers for example, transfer information at relatively low rates, on the order of 100 to 150 words per minute maximum. If one assumes each word is an average of six characters long and each character requires eight bits to transmit, then the effective data rate is only about 120 bits per second. If the system uses ordinary 25 kHz dispatch frequency slots, then the spectral efficiency as measured in bits-per-second per Hertz of bandwidth is less than.005.

However, using modern techniques, it is now easy to send 4,800 bps in a 25 kHz channel and much higher rates are achievable — as high as 25 kbps or more. Stated another way, even at 4,800 bps, sending information as data messages rather than voice messages increases the throughput and spectral efficiency by a factor of 40. At 25 kbps, the improvement could be as high as 200. Heavy users of mobile communications, overnight courier services for example, take advantage of this by loading several hundred mobile data terminal users on a single channel — many more than can be accommodated using voice on a single channel.

Two-way voice dispatch and cellular mobile radio systems can be used for transmitting data messages by using modems and a "dial-up" connection. However, since these systems were optimized for voice communications, they are subject to limitations when used for sending data. Not only must the modem contend with a much more hostile channel with time varying characteristics such as multipath distortion, signal fading, noise, and interference, but also with mobile/portable customer equipment that contains special circuitry that enhances voice performance but constrains or complicates data handling. In addition, in an analog cellular mobile radio system, certain signaling functions are carried out over the voice channel itself. This is done by briefly interrupting the voice signal and sending a burst of digital signaling information. Such a burst of digital signaling is used, for example, to order the mobile unit to change channels during a handoff from one cell to another. The interruption is so short that it typically remains unnoticed to a voice user, but it can cause havoc with ordinary dial-up modems. Modem manufacturers have responded to the growing interest in using analog cellular systems for data communications by developing modems containing special protocols aimed at overcoming or at least minimizing these problems. In addition to these problems, there are often difficulties associated with connecting the modem to the mobile/portable unit. This is because the physical and electrical interfaces to the mobile/portable unit are not standardized and, in some cases at least, may not even contain desired electrical connections.

In any event, the use of circuit switched systems is not efficient for handling "bursty," computer data where the integrity of the data is critical to the service or application. In this part of the report, the focus is on specialized systems that have been optimized for the provision of two-way data services. A number of companies operate such specialized two-way mobile data systems either for their own private use or for the provision of service on a third-party basis. In certain respects, these two-way data systems can be looked at as extensions of the specialized packet switched networks that have evolved using terrestrial transmission facilities.

In these specialized systems, the message (i.e., collections of alphanumeric characters) to be sent is divided up into smaller units of information called packets. Each packet includes the address of the person or equipment to whom the message is directed as well as extra bits that are used to detect (and perhaps correct) errors in transmission. In contrast to circuit switching used for voice services, in packet switching there is no full-time direct connection between the two ends of the connection. Instead, these packets are sent from node-to-node through the network in store-and-forward fashion and, at the receiving end, the individual packets are reassembled into the original message.

There are a number of advantages to using this technique. First, because there is no "realtime" connection, each packet can be disassembled at each node and checked for errors. If an error has occurred, it can be corrected based upon the extra bits transmitted or by requesting the forwarding node to resend the packet that is in error. In this way, error free messages can be assured. Second, packets from different users can be commingled on the same link; that is, multiple users can share the same channel. If one unit has nothing to send, it transmits no packets and no channel capacity is used. When it does have something to send, its packets are interleaved with the packets being sent by other terminals. This technique is particularly well suited for "bursty" data traffic because capacity is only consumed when information is being sent.

The tradeoff with packet switching, of course, is that there are delays associated with getting on to a link and with disassembling the packet, checking and correcting errors, and reassembling the packet at each node. These delays make traditional packet switching unsuited for voice. However, delays of several seconds are perfectly acceptable in many data applications (e.g., in notifying a central computer that a package has been picked up). From a user standpoint, this delay is often more than offset by error free performance and increased efficiency in the use of network resources.

In a packet switched, two-way mobile data system, the region to be covered is divided up into small areas or cells, each served by a base station transceiver and associated antenna equipment. These cells are then interconnected to the fixed network using wireline or other facilities. The over-the-air signals are sent digitally (as described before) and, because channels with paired frequency slots are utilized, full duplex (i.e., simultaneous transmissions in both directions) is possible.

The mobile units and base stations exchange packets comprising the messages as just described. In the in-bound direction, the base stations receive the packets and, based upon the address, route them toward a destination such as the customer's computer. The packets may be picked up by more than one base station, but the duplicate packets can be discarded using logic in the network. Similarly, in the out-bound direction, the mobile unit receives packets that are addressed to it and discards the rest. Error free performance over otherwise marginal radio links can be assured using the error checking/correcting techniques just described.

Another interesting advancement in the provision of two-way mobile data services is the development and deployment of the Cellular Digital Packet Data (CDPD) system by portions of the cellular industry. These systems, which — as the name implies — use the packet data approach just described, are overlaid on existing analog cellular networks in such a way that they can utilize a portion of the existing cellular infrastructure. The packetized data transmission occurs either during normal gaps in the usage of the cellular channels for voice communications or on separate cellular channels that have been set aside for such purposes. In either case, the normal analog cellular operations can continue without interference or interaction with the data traffic.

The CDPD system allows the carrier to exploit economies of scope or integration while providing customers with a wide coverage, packet data communications capability or a combination of normal cellular voice services and packet data services within a single device. A large number of cellular carriers support CDPD.

Two-way Dispatch

Major developments in the systems for providing two-way dispatch have occurred in the multichannel trunked radio systems described in Section III.E. As described there, compared to conventional repeaters, the early trunked systems offered greatly improved dispatch services by providing disciplined access to channels and much greater spectral efficiency. Subsequent developments in trunked systems include the use of more powerful signaling techniques, virtually unlimited addressing capabilities, smooth integration of dispatch and interconnected mobile telephone service, and multi-site networking capabilities.

The combination of greater addressing capabilities and more powerful signaling, for example, allows the creation of different communications groups among the units that comprise a customer's fleet. Thus, a dispatcher at a company that owns cement trucks and sand and gravel trucks can, at his or her option, communicate with the entire fleet, subfleets consisting of just the sand trucks, gravel trucks, or field supervisors, or, on a private basis, with individual units. Moreover, using high-speed signaling, the network can be dynamically reconfigured, thus creating what is often referred to as a software defined network. The ability to reconfigure networks dynamically is especially useful in emergency or other abnormal circumstances where field resources need to be redeployed to fit a particular circumstance.

Another major trend in trunked radio systems, like in telecommunications networks more generally, is the movement toward digital rather than analog transmission. Digital transmission has a number of advantages such as, for example, the ease of providing different services on a common platform. But a principal impetus for the use of digital rather than analog transmission in radio systems is that, when it is combined with voice compression and more efficient modulation techniques, it allows multiple voice conversations to be carried in the same frequency slot that formerly carried a single analog voice conversation. Typically this is done using the TDMA techniques described earlier, although FDMA and CDMA techniques are being proposed/deployed as well. One commercial TDMA-based system now being introduced in the United States allows six voice conversations to be carried simultaneously in a single, 25 kHz wide frequency slot or channel.

Another, somewhat less clear trend in trunked systems is toward more cellular-like configurations with smaller coverage areas for each base station. For one-to-one communications (e.g., mobile telephone service), this allows greater spectrum efficiency through increased frequency reuse. However, this approach is problematical for one-to-many (fleet dispatch) communications because it requires, in effect, separate calls to be simultaneously set up to each cell which contains a unit in the fleet. This may reduce both the technical and economic performance of the network in the dispatch mode. Thus a network configured with large cells provides better performance in fleet dispatch applications at the penalty of low spectrum efficiency in mobile telephone applications, while a network configured with small cells provides better spectrum efficiency (and hence greater capacity) in mobile telephone applications, but at the penalty of poorer performance in dispatch applications.

Two-way Mobile Telephone

Major developments in systems for the provision of wireless telephony have occurred on four fronts. First, in response to reduced regulatory restrictions and for other reasons, trunked, two-way dispatch systems have improved significantly in their ability to provide wireless telephone services. Second, due to rapid growth in the number of subscribers, evolving competition and other factors, the cellular systems operating in the 900 MHz region of the spectrum in the U.S. have changed significantly in technological terms. Third, the FCC has reallocated and assigned (through the use of auctions) a large block of new spectrum in the 1.9 GHz region for the provision of Personal Communication Services — including mobile/portable telephone services. Fourth, there have been significant technological developments in cordless telephones and related devices using unlicensed spectrum — including additional unlicensed spectrum in the 1.9 GHz region.

The developments in trunked two-way dispatch systems that allow them to provide wireless telephone services more effectively were described in Section IV.C, immediately above, and will not be addressed further here. Developments on the remaining three fronts are described in the paragraphs which follow.

Cellular

The most far-reaching trend in cellular mobile radio systems is the conversion from analog to digital transmission. In the U.S., the original AMPS systems operating in the 900 MHz band used FM modulation in frequency slots that are each 30 kHz wide. The standard that is associated with AMPS is known formally as IS-553. Thus, in AMPS, each direction of a two-way conversation uses a 30 kHz-wide slot — a width that is almost ten times wider than the voice signal that is being transmitted.

The phenomenal growth of the cellular service, coupled with the difficulties and expense of moving to extremely small cells in order to increase capacity through frequency reuse, has put pressure on the service providers and their equipment vendors to develop more spectrum efficient transmission methods. Digital compression and transmission techniques were chosen as one method of achieving the needed increase in capacity. The situation was complicated by the need to maintain backward compatibility with the existing analog systems during at least a transition period, if not indefinitely.

Two incompatible digital standards have been adopted for the conversion of existing cellular systems to digital techniques in the U.S. The first divides existing frequency slots/channels using TDMA, while the second divides a contiguous set of channels (i.e., a block of existing cellular spectrum) using CDMA. The standard for the former, which is based on providing three time slots (and hence three simultaneous voice conversations) in each channel is known as IS-54 and the latter, for which proponents claim even greater capacity gains, is known as IS-95. Both standards are "dual mode" which means that a handset is capable of operating in either the digital mode or in the analog mode.

This means that a customer using one of the new dual mode subscriber units can operate on either an analog system or a digital system. Thus, if the customer roams to a system (in a rural area, for example) that has not converted to the digital standard, he or she can still place and receive calls. Likewise, a cellular carrier can operate a mix of analog (IS-553) and digital (either IS-54 or 95) systems in order to continue to provide service to users who have not purchased digital subscriber units while still achieving greater capacity on the digital channels. Over time, as the percentage of analog-only subscriber units decreases, the percentage of channels devoted to the less spectrum efficient analog technology can be reduced. Note that, because the industry has not settled on a single, compatible digital standard, the analog channels are necessary to provide "a lowest common denominator" for roaming customers. Thus, in the absence of a common digital standard (or the development and full deployment of dual-mode TDMA/CDMA subscriber units), all providers must continue to provide analog support channels.

Besides the over-arching trend toward digital transmission and TDMA and CDMA as alternative access techniques, there are several developments worth mentioning. First, the industry has been able to develop standards that, among other things, facilitate handoffs between systems and the provision of automatic roaming. Automatic roaming, of course, refers to the ability of the customer to place and receive calls automatically as he or she moves across cellular systems owned by different providers and utilizing infrastructure equipment manufactured by different vendors. These developments are based, in part, on increased reliance on advanced signaling techniques associated with the emergence of digital transmission standards and with inter-switch communications based on common channel signaling.

Second, vendors have developed and providers have deployed microcells in order to provide better coverage inside buildings and other difficult to serve locations. Such microcells become increasingly important as low-power portable units (rather than higher power mobile units) proliferate and competition from new Personal Communications Systems operating in the 1.9 GHz band emerge. On a related point, cellular carriers and their equipment vendors have also developed handsets that operate on unlicensed cordless telephone frequencies in one mode and as ordinary cellular handsets in another mode. This means that the customer can use the unlicensed mode to place and receive calls without incurring airtime charges when in the vicinity of the cordless telephone base station (say, inside an office complex), while still having full cellular service when outside that range.

PCS

As mentioned earlier, one of the most important recent developments impacting on the provision of wireless telephone services in the U.S. is the reallocation of a large block of new spectrum for the provision of Personal Communications Services or PCS. One of the difficulties of dealing with PCS is that it has no single, clear definition. However, the ultimate concept or vision of a Personal Communications Service seems to revolve around the notion of the ability of customers to obtain personalized telecommunications services (e.g., to make or receive calls) at virtually any location at virtually any time. The essence of PCS seems to be a wireless service utilizing a common handset and a single telephone number (referred to as Personal Number Calling). Under this latter concept, customers would receive a single telephone number and that single number would no longer be associated with a physical location on the PSTN. Rather, intelligence residing with the network would be used to route calls to the individual regardless of his or her location.

It should be emphasized that PCS encompasses not a single network, but a seamless web of interconnected networks that, as one observer defined it, is "an extension of and integration of current and emerging wired and wireless telecommunications network capabilities allowing communications with persons and supporting personal mobility." This vision of PCS is broad enough to encompass today's paging and cellular services, but the focus in this section of the report will be on the development, deployment and evolution of PCS in the newly reallocated spectrum in the 1.9 GHz region.

In the 1.9 GHz region, the FCC has reallocated a total of 140 MHz for PCS. It has divided this spectrum up into six licensed blocks of differing sizes, and they have labeled the blocks with the letters A through F. Blocks A, B, and C each contain a total of 30 MHz, while blocks D, E, and F contain 10 MHz each. Each of the blocks is split evenly into two parts separate sufficiently in frequency to allow full-duplex operation as explained previously. All six of these blocks, which total 120 MHz, are for licensed services. The Commission's rules allow the 10 MHz blocks to be combined with the 30 MHz blocks to create a maximum of 40 MHz of spectrum for one system. The remaining contiguous block of 20 MHz has been set aside for unlicensed operation and it has been further divided into two parts — one for voice services and one for data services.

The technical rules that the FCC has established for the use of the newly reallocated spectrum for licensed PCS services are loose enough and non-specific enough to accommodate different types of PCS systems. Out of this flexibility, two basic types of PCS systems seem to be evolving. The two types are often referred to as "low tier" (or "low mobility") and "high tier" (or "high mobility").

The low tier systems can be thought of as evolving from today's cordless telephone technology and market. Like cordless telephones, these systems would provide wireless access to today's wired network while the user is on the customers' premises, but the same handset would also provide access to the wired network away from the home, office or factory. More generally, these low tier PCS systems are expected to be deployed in cases where low mobility is involved, wireline telephone quality voice is needed, customer usage approaches local loop usage levels, traffic density is high, low cost (suggesting low complexity) is desired, and service to low cost, light weight handsets with long battery life is required. Low mobility means that the system is only capable of handing off calls at relatively low speeds — i.e., at walking speeds.

The high tier systems can be thought of as evolving from the existing cellular mobile radio technologies. In fact, many observers regard the developing high tier PCS systems as being cellular systems that operate in another region of the spectrum — namely 1.9 GHz rather than 900 MHz. In keeping with the similarity between today's cellular and the evolving high tier PCS systems, the latter are expected to provide service where high mobility is involved, voice quality is somewhat less of a requirement, customer usage levels are more typical of today's cellular rather than ordinary wired telephones, traffic density is less, and the size, weight and cost of the handsets are somewhat less critical.

These different requirements for low tier versus high tier systems result in different network architectures or system designs. It is beyond the intended scope of this report to describe the various technical tradeoffs in detail, but a few of them will be suggested. The higher voice quality requirements for the low tier system, along with the need for low cost, light weight handsets, implies the use of less voice compression — i.e., the use of waveform encoding and decoding. As described earlier, higher bit rate speech not only improves quality, but it also reduces the amount of computer processing power employed in the handset. A reduction in processing power, in turn, translates into a reduction in battery drain and a corresponding increase in battery life. On the other hand, higher bit rate voice is less efficient in the use of the radio spectrum resource and that, coupled with the need to provide service to very low power handsets in high usage, high traffic density situations, translates into the need for cells with much smaller radii than traditional cellular systems. Consequently, smaller cells, typically on the order of.25 miles or so, are needed both to increase the amount of frequency reuse obtained in the area and to allow service to the low power handsets. With such small cells, cell-to-cell handoffs become extremely difficult and work against the need to keep complexity and costs low. Thus, as explained above, low tier systems, which may compete with wired telephone systems, may only offer handoffs at pedestrian speeds. In addition, the small cells may make it less practical to provide seamless coverage over less densely populated areas.

The larger cells associated with traditional cellular systems allow the coverage of larger areas with fewer base stations and ease the handoff problem. But larger cells reduce the amount of frequency reuse obtained and imply the need for higher power handsets. The reduced spectrum efficiency can be compensated for by an increase in the amount of voice compression used, but at the expense of voice quality and battery life. However, the resulting voice quality may be more than adequate since the handsets are often used in vehicles and other noisy places where the extra quality is not noticeable or critical. Likewise, access to the vehicle's battery lessens the need to depend entirely on the battery in the handset.

These and other tradeoffs interact in complex ways with market considerations, and the details of how all of these tradeoffs will be made is far from certain at this point. However, two contenders have emerged for low tier system standards, four for high tier, and one for a "middle tier." One of the contenders for the low tier category is a system known as PACS. It is a combination of a system developed at Bellcore known as WACS and the Japanese Personal Handiphone (PHP) system. It uses TDMA, frequency division duplexing, a form of phase modulation, 32 kbps voice coding, and handsets with an average transmitter power of 12 milliwatts. The second low tier contender is based on the Digital European Cordless Telecommunications (DECT) standard. It uses TDMA, time division duplexing, a form of frequency shift modulation, 32 kbps voice coding, and handsets with an average transmitter power of 10 milliwatts.

One of the four contenders for the high tier systems includes a system based on the U.S. TDMA digital cellular standard (IS-54). It employs TDMA as the access method, frequency division duplexing, a form of phase shift modulation, approximately 8 kbps voice coding, and handsets with an average transmitter power of 200 milliwatts. A second contender for the high tier system is based on the U.S. CDMA digital cellular standard (IS-95). It employs CDMA as the access and modulation method, frequency division duplexing, variable voice encoding (8 and 13 kbps), and handsets with an average transmitter power of 200 milliwatts. The third system is based on the second generation European digital cellular standard known as GSM. This system, sometimes referred to as DCS-1900, employs TDMA, frequency division duplexing, a combination of phase/frequency modulation, 13 kbps voice coding, and handsets with an average transmitter power of 125 milliwatts. The fourth contender is also a CDMA system, but one which spreads the signals over a wider bandwidth. The single "middle tier" contender employs a combination of CDMA and TDMA as the access method, time division duplexing, a range of voice coding speeds, and handsets with an average transmitter power of 62 milliwatts.

It is almost certain that different standards for PCS will emerge in the U.S., not only between low tier and high tier systems, but within the two categories as well. What systems will actually be deployed and whether the industry will eventually converge on a single standard (or, perhaps, single standards for high tier and low tier systems) is still uncertain.

Unlicensed Two-Way Voice

It was pointed out that today's unlicensed analog cordless telephones have proven enormously popular despite rather severe limitations in the number of channels available. The FCC recently expanded the number of channels available in the 49 MHz region of spectrum from ten to 20 and much more sophisticated cordless telephones using spread spectrum techniques have developed. The latter cordless telephones operate in the 900 MHz region of the spectrum. Recall that it is possible to construct handsets that include both cellular and unlicensed cordless telephone technology. As pointed out before, this allows the customer to avoid air time charges when he or she is at home or in the office.

Some manufacturers have extended the single line, cordless telephone technology to create wireless Key Telephone Systems (i.e., multiline "push button" telephone systems) and wireless Private Branch Exchanges (PBXs) for business applications using the unlicensed spectrum. These systems use cellular techniques, but with very small cells scattered over the office, factory, or campus to provide good coverage and higher capacity. These systems allow employees to make and receive telephone calls while away from their desks or other regular location and reduce the costs of the "moves and changes" that are associated with moving employees around due to growth or reorganizations.

The reallocation of an additional 20 MHz of spectrum for unlicensed uses in the 1.9 GHz PCS band will permit further expansion of wireless voice services, including wireless KTSs/PBXs. Of the 20 MHz of spectrum, the Commission has allocated 10 MHz for voice services as reported earlier. Systems in the new band will utilize a special industry developed "Listen before you talk" etiquette to avoid interference between nearby systems. Since the unlicensed band is in such close proximity to the licensed bands, it will facilitate the development of handsets that will operate in both the licensed and unlicensed portions of the PCS spectrum.

INTERRELATIONSHIP BETWEEN WIRELINE AND WIRELESS DEVELOPMENTS

One of the most important trends in the development of the wireline telephone network is movement to common channel signaling, or separating voice traffic from signaling traffic. Signaling involves the exchange of information associated with setting up, maintaining and taking down calls. In the wireline network, up until fairly recently, analog tones were used to convey signaling information which was carried on the same channel as the voice conversation. With common channel signaling, all of the signaling information associated with multiple conversations is digitized and handled on a common packet-switched subnetwork. Common channel signaling allows calls to be set up faster because the signaling itself is faster and more efficient since signaling does not consume conversation capacity on the circuit switched voice subnetwork. The specific common channel signaling system now being widely deployed is referred to as Signaling System 7 (SS7).

Another major trend in the wireline network is toward unbundling of stored-program control switching systems. Modern computerized or stored-program control circuit switches are composed of two basic parts — the matrix where physical connections are made between circuits and a processor that contains the logic that controls the switching. In early stored-program control switches, the switch (matrix) and processor elements were tightly integrated, and this tight integration ultimately caused a number of disadvantages. The modern trend of separating the switch control from the lower level switching functions overcomes these disadvantages and allows entire networks to become programmable.

These advances in signaling and switching allow the creation of what is referred to as the "Intelligent Network." Computer processors and their associated databases are placed in the network where they can be accessed from the signaling channel. The system uses the calling and called numbers and other information to handle calls in special ways — e.g., to route calls to different locations depending on the time of day and/or the originating location. One obvious application of these intelligent network capabilities is in providing customers with the ability to roam among different locations and even among different networks. For example, a switch can suspend a call briefly, look up the called number in a database, and then route the call to the intended recipient at another number and location on the same system (call forwarding). Similar techniques can be used in wireless networks to allow customers to roam from system to system. When a wireless customer carrying a portable telephone arrives in a distant city, for example, the visited system recognizes that the user is roaming and automatically notifies the user's home system that the handset is in the new area. This information is placed in a database called a home location register. Subsequently, when a call is placed to the user's number on the home system, the call is automatically routed to the customer in the distant city.

Due to regulatory as well as technological constraints, early wireless systems were built entirely separate from the wireline network, and it was only later that interconnection became possible between the two. This separateness carried over to early cellular radio systems but, more recently, there has been increased interest in the issues associated with interconnecting and even sharing the databases, processors, and common channel signaling associated with wireline and wireless networks of all types. Such seamless interconnection/resource sharing will allow the efficient provision of powerful new services such as personal number calling. With personal number calling, customers no longer have to have multiple telephone numbers for their home, office, cellular phone, fax machine and other lines, but rather they receive a single number that is not associated with a fixed, physical location on a network. Calls to that number will then be routed to the customer wherever they are and irrespective of the network by which they are currently being served. This will be done by using the databases, processors and signaling capabilities comprising the Intelligent Network to route the call. Thus there is a significant interrelationship between wireline and wireless developments in supporting the concept of personal mobility described earlier.

Unlicensed Two-way Data

Broadly speaking two-way wireless data communications can be divided into two categories. The first is the wide-area data systems of the type described in Section IV.B. The second category is local data systems that provide service over a more limited area such as an office, factory, or campus. These local systems are referred to as wireless Local Area Networks (LANs). Wired LANs have grown extremely rapidly in recent years as a way of interconnecting personal computers for, among other things, the exchange of electronic mail, the sharing of peripheral equipment such as printers and high-speed modems, and accessing large centralized files.

Wireless LANs, like wireless KTSs/PBXs in the voice world, allow more mobile employees (e.g., warehouse workers) to be connected into the organization's wired network and reduce the costs of moving employees from place to place within the business complex. There are a number of commercial products available today that operate in the unlicensed band in the 900 MHz region as well as other unlicensed bands near 2.4 and 5.7 GHz. Typically these systems employ packet techniques and some form of spread spectrum as the modulation and access technique. Efforts are now underway to create standards for wireless LANs.

Because of the wider bandwidths available and the frequency reuse that is possible with the short ranges involved, much higher data rates are possible than on the wide area, two-way wireless data systems described previously. As noted earlier, the FCC has allocated 10 MHz of spectrum in the 1.9 GHz region for unlicensed data applications and, due at least in part to certain interference issues in that band, has indicated a willingness to reallocate other spectrum for wireless LAN and similar applications.

Mobile Satellite for Voice and/or Data

This report has focused primarily on terrestrial based systems, and space does not allow a full discussion of satellite based systems. However, it should be noted that satellites in geostationary orbit have been used for many years to provide voice and data services to mobile terminals and, more recently, organizations have moved to deploy constellations of low earth orbiting (LEO) satellites that are suitable for communicating with relatively small, low power, handheld, subscriber equipment.

Thus, one can envision the day when a customer can obtain service virtually anywhere in the world using a common handset (or a different handset using a form of "smart card" associated with the particular user). Such a system would automatically select the lowest cost service that delivers the desired quality. So, for example, the user could access an unlicensed cordless telephone-like system when at home or at the office, a public low tier PCS system when in high density locations, a high tier PCS system (including cellular) when in a vehicle and/or less urban location, and a satellite-based system when in remote locations outside the range of all terrestrial-based networks.

EVALUATION OF THE MEDIUM TERM TECHNOLOGICAL AND SERVICE TRENDS

The previous two subsections (IV.A and IV.B) provided, respectively, an overview of advances in enabling technologies and a description of short to medium term wireless telecommunications technology and service trends that take advantage of those advances in response to user demands and competitive marketplace pressures. The purpose of this subsection is to summarize the medium term trends and to evaluate them in terms of their implications for people with disabilities.

Summary of Medium Term Trends

Broadly speaking, the trends identified earlier can be divided into five categories — improved coverage, integration of services, increased network-based functionality, increased functionality of end user equipment, and improved spectrum efficiency. These broad trends are discussed in more detail in the paragraphs which follow.

In evaluating the coverage of wireless telecommunications systems, it is common to speak in terms of both the breadth and the depth of coverage. The breadth of coverage refers to the outer limits of the geographic area over which service can be obtained. Depth of coverage refers to the ability to receive service at locations with those outer limits, e.g., within buildings, tunnels, and similar hard to reach locations. Because of the technological developments described earlier, competitive pressures, regulatory build out requirements, and other factors, both the breadth and depth of coverage of wireless telecommunications systems have increased significantly in the United States over the past decade. For example, today's cellular mobile radio systems provide coverage over 70 per cent of the land area which translates into population coverage approaching 96 per cent. Moreover, what were once independent wireless telephone, dispatch, paging, and two-way wireless data systems are being knit into larger networks that facilitate universal roaming. Intensifying competitive pressures and the development and deployment of new microcellular systems at one extreme and mobile satellite systems at the other, holds out a strong promise for the continuation of these trends toward improved coverage and universal roaming.

For a host of regulatory, technological, and economic reasons, wireless telephone, dispatch, one-way paging, and two-way data services in the past were supplied on separate networks. However, there is apparently a strong demand from users for integrated combinations of services. For example, some users may desire a combination of one-way paging and dispatch service, while another may want voice dispatch and mobile telephone service.

Fundamentally, there are two ways of responding to this demand for integrated services. A provider can offer multiple services on a common network or furnish the customer with end user equipment that is able to receive different services from separate networks. In other words, the integration can be provided either within the network or within the handset or other end user device.

The former solution, integration within the network, has the advantage that it helps capture any network-based economies of scale or scope. Such economies can be important in networks that require the construction of expensive backbone networks over vast areas. On the other hand, because of the fundamental differences between the requirements for voice dispatch, wireless telephone, paging, and two-way data services that were discussed earlier, there are inevitable engineering compromises that arise in creating a "one size fits all" network. This may lead to a network that does not provide any one of the multiple services as effectively or efficiently as a standalone network dedicated to the provision of a single service. In such situations, it may be more effective to provide the desired integration within the end user equipment allowing the construction of separate networks to capture the economies of specialization associated with standalone networks customized to provide a single or more limited set of services.

Examples of both types of integration were provided earlier. Modern trunked radio systems have been designed to provide both dispatch and mobile telephone services on a common platform, and packet switching equipment has been added to cellular mobile radio services to render them more effective in providing two-way wireless data services. Modern cellular/PCS systems often combine mobile telephone service with a "short message service" which is similar to one-way, alphanumeric paging. These are examples of the service integration being provided on a common network. Combining an unlicensed cordless telephone with a cellular or PCS wireless telephone in a single handset is an example of providing integrated services on separate networks with the integration occurring in the handset. Combining a cellular mobile telephone unit with a satellite mobile telephone unit in a single mobile unit is another example. Given marketplace demands, the advances in enabling technologies described earlier and the increased competitive pressures associated with new entrants into the wireless market, it is likely that these trends toward integrated service provision will continue in the medium term.

In basic terms, functionality reflects how well equipments or systems are designed or adapted to suit a particular function or use. From a wireless end user perspective, improvements in functionality can arise from two fundamental sources: improvements in the network and improvements in end user equipment. Other than the improved functionality associated with the greater breadth and depth of service coverage of wireless systems already described, most increases in wireless network functionality can be attributed to the increased incorporation of computer processing power with the network itself. As pointed out earlier, this increase in network based intelligence, coupled with more powerful signaling techniques, can be used, for example, to allow customers to dynamically reconfigure networks, thus allowing the creation of software defined networks. It is this same general concept that allows implementation of the personal number calling and enhanced levels of personal mobility — all examples of network based improvements in functionality. With increased competition generated by new entrants in the wireless telecommunications marketplace, it seems safe to predict that over the medium term competitors will continue to use network based intelligence to create customized services in order to differentiate their services from those of their competitors.

Improvements in battery technology and higher levels of integration in digital integrated circuits have allowed cellular subscriber equipment to shrink in size to the point where units that will slip into a shirt pocket are now available. Increased digital processing power in the end user equipment (coupled with improved signaling techniques between the network and that equipment) has allowed end user equipment, for example, to assist the network in making handoff decisions in mobile telephone systems, thus improving the performance or functionality as seen by the end user. The technical (and economic) feasibility of allowing more processing power to be placed in the end user equipment facilitates improved functionality. Examples include the storage of frequently called numbers in cellular handsets and, more recently, the storage of voice messages in paging receivers. In essence, this means that the end user equipment can not only improve the basic performance of the network, but also be customized to meet the specific needs (e.g., working habits) of the customer. Improvements in display technology (e.g., in liquid crystal displays) allow numerical, text, and image messages to be offered visually. The same display technology can also be used to help the end user to "reprogram" the device to fit his or her needs. With some exceptions largely associated with proprietary systems, there are many more suppliers of wireless end user equipment than wireless infrastructure equipment and, accordingly, the market is highly competitive. While there are obviously some practical limits to how small subscriber units can be made and still be useful in certain applications, this intense competition coupled with the continued technological advances in digital integrated circuits, for example, means that the trends in improved functionality in end user equipment are apt to continue in the medium term.

Developments such as source coding and greater frequency reuse that lead to improved spectrum efficiency in wireless systems were discussed earlier. These developments, coupled with significant reallocations of spectrum for terrestrial- and satellite-based land mobile radio systems as a result of regulatory actions, suggest that spectrum scarcity may not seriously impede the further development of land mobile radio services (broadly defined) over the medium term.

Evaluation of the Medium Term Trends

With some exceptions noted later, the medium term trends in wireless telecommunications technology and services would appear to bode well for not only the general public, but disabled Americans as well. Like all consumers, Americans with disabilities will benefit from the improvements in the breadth and depth of service coverage that are being driven by competitive pressures and by technological advances ranging from advanced cordless telephone systems to the latest geostationary and low earth orbiting mobile satellite systems. Perhaps the most favorable trend for the disabled is the movement toward the provision of voice (both dispatch and mobile telephone), data, and image services on an integrated basis through integration in the handset, integration in the network, or both. This is a very favorable trend because a key concept in providing access to information technologies to persons with disabilities is to deliver the information in the most accessible form — e.g., in text or image form for the deaf and in auditory or tacile form for the blind.

Moreover, because the wireless telecommunications networks themselves are becoming increasingly intelligent and programmable, it is increasingly possible for the network itself to respond to any special needs of the disabled subscriber. For example, because the network is, in effect, becoming increasingly programmable, it can (a) recognize from the address and/or service profile of the customer being called that the information should be delivered in a particular form and (b) change or translate the information (say from text to voice) in order to be able to deliver it in the most accessible form. It is important to note that this ability for the network to, first, recognize that the consumer's desire to have information in a particular format, to, second, make the translation in format if necessary and to, third, deliver the information in that format may have high value even for customers who are not disabled. For example, a sighted person driving to work may prefer to have waiting E-mail messages (or at least the "from" and "subject" lines) delivered by voice to his or her cellular phone or pager, but, once at work, prefer to have them delivered to his or her personal computer connected to the office local area network.

The above illustrates an important principle long recognized by advocates of universal accessibility; namely, making networks or other facilities more accessible to the people with disabilities makes them more accessible for everyone. And, even better, the advancing technology and falling costs trends for wireless telecommunications networks and services reported herein mean that the "win-win" situation associated with universal accessibility can be more readily achieved now and in the future.

While the medium term trends in wireless telecommunications technology and services appear to bode well for not only the general public, but for disabled Americans as well, there are concerns in terms of universal access. Four particular concerns will be discussed in the following paragraphs:

First, in order to gain the benefits of more intelligent networks and terminals, they must be programmed or customized often by the subscribers themselves. Simple examples include instructing the network to forward incoming calls to a particular number during a certain part of the day or reprogramming an alphanumeric pager to deliver notification of an incoming text message by vibrating rather then beeping. In order to make these programmable features and functions accessible to the disabled, good design practices must be followed. For example, input-output devices that require good eye-hand coordination should be provided with alternative or supplemental modes which do not require such coordination. Note, once again, that reducing the eye-hand coordination required increases accessibility for everyone. Second, the rapidly decreasing size of end user equipment (e.g., cellular telephones and pagers) may create accessibility problems for certain users. In other words, the competitive race to produce even smaller and lighter end user equipment inevitably puts pressure on the designer to decrease the size of keyboards, visual displays, and other built-in input-output devices.

On a related point, the size, emphasis on battery life, and overall design of end user equipment such as cellular telephones can also create obstacles for hard of hearing or deaf people. These obstacles include difficulties in successfully coupling their hearing aids to the subscriber unit or in utilizing acoustic couplers to connect to their text telephones. In the case of text telephones, the strong emphasis on small size may also discourage manufacturers from providing built-in jacks to allow direct connection of such devices. The disabled, like all consumers, can benefit from the convenience associated with such decreases in size and weight of end user equipment, but if accessibility for, say, those with physical impairments cannot be provided directly as a result, then provisions should be made for the use of supplementary devices to accomplish the same purpose.

Third, one clear trend in both wired and wireless networks is a shift from analog to digital transmission. There are a host of benefits associated with this shift and it is these benefits that are driving the shift in both segments of the industry. In the wireless portion of the telecommunications industry, the trend has been coupled with digital voice compression. As explained in Section IV.A.3., the source coding used in voice compression systems uses computer processing power to remove redundancy from speech so that (a) fewer bits per second have to be transmitted to convey a voice signal and (b) the available bandwidth can be used more efficiently. Most operators of commercial wireless telephone systems have a strong incentive to employ digital voice compression because the lower bit rates translate into (a) more conversations in a given amount of spectrum — i.e., more efficient use of the radio spectrum and greater capacity, and (b) more conversations per piece of radio equipment and/or radio site — i.e., greater economies of scale. An unintended side-effect of employing voice compression is that it can cause problems for non-voice signals such as those emitted by fax machines, computer modems, and, especially important in the context of this report, text telephones. This is because the vocoders depend critically upon the input signal having the characteristics of the human voice.

There are a number of potential solutions to this problem, including techniques for bypassing the voice compression devices and thereby gaining direct access to the digital bit stream being transmitted or received. It is beyond the scope of this report to describe these techniques in detail, but it should be noted that such techniques are being developed to facilitate the high-speed transmission of fax and computer communications on a circuit switched basis as well. As engineers and their managers improve the data handling capabilities of their emerging digital networks, it is important — indeed imperative — that they also consider the needs of text telephone users.

Fourth, in the public policy arena, there is a strong and clear trend toward the FCC and other regulatory agencies becoming less involved in developing and promulgating standards in the wireless industry. This trend reflects, among other things, growing concerns about the time consuming nature of formal standards-making activities, the ability of the government — no matter how well motivated or well intended — to pick an appropriate standard in the face of rapid technology and market changes, and the tendency of standards to sometimes inhibit rather than promote technological advancements. This trend toward relying more upon private, voluntary standards and less upon governmentally mandated standards has a number of implications relating to the accessibility and usability of wireless telecommunications to people with disabilities. For example, instead of dealing with a single government agency in a formal proceeding addressing the establishment of a specific common standard, the disabled community will, most likely, have to participate in broader agency proceedings of more general applicability. At the same time, they will have to work with multiple private sector providers each using different technologies and proposing different solutions for improving accessibility and usability. This discussion illustrates that one unintended consequence of increased reliance on voluntary standards is that Americans with disabilities and the organizations that represent them will face a greater challenge in assuring such accessibility and usability even with the support provided by recent legislation.

Thus, while the medium term trends in wireless telecommunications technology and services generally bode well for disabled Americans, there are significant concerns that must be addressed if universal accessibility is to be achieved.

REGULATORY TRENDS FOR THE FUTURE DEVELOPMENT OF WIRELESS TELECOMMUNICATIONS

The purpose of this section of the report is to briefly review the legislative and regulatory trends that will influence the future development of wireless systems and services. It is beyond the scope of the report to examine each legislative and regulatory action in detail and, besides, the regulatory proceedings are moving at such a pace that the discussion would become quickly outdated. Instead, the broad trends in policy and regulation will be described and examined in terms of their implications for future developments in radio-based systems and services.

The general trends that can be identified include (a) the reallocation of spectrum from Federal government use to non-Federal government use, (b) the allocation of more spectrum for mobile as opposed to fixed applications,

(c) the use of auctions to assign spectrum to particular users, (d) increased licensee flexibility in the use of assigned spectrum, (e) continued support for unlicensed services, (f) increased competition in the provision of all telecommunications services, including radio-based services and (g) increased reliance on voluntary standards. Under the Communications Act of 1934 (as amended), the FCC only has jurisdiction over non-Federal government spectrum. Jurisdiction over spectrum licensed to Federal agencies resides in the Executive Branch. As a result of legislative actions taken in recent years, substantial blocks of spectrum have been transferred from Federal government use to private and commercial uses under the jurisdiction of the FCC. Arguably, the motivation for this transfer stemmed from (1) the belief that the Federal government did not need all of the spectrum that it controlled and (2) the interest in collecting additional revenues by auctioning the transferred spectrum. In any event, the result has been a substantial shift of spectrum to the FCC.

In addition to more spectrum being made available for commercial and private uses, the FCC has taken steps to increase the amount of spectrum available for mobile radio use. For example, the FCC released a significant amount of additional spectrum for narrowband PCS (i.e., one-way and two-way paging systems) and reallocated spectrum previously used for fixed point-to-point microwave services to broadband PCS as discussed earlier. Moreover, as part of its proceeding dealing with the Advanced Television Service (or Digital Television — DTV — as it is now referred to), the FCC recently suggested the possibility of reallocating television broadcast spectrum to land mobile radio use. It is likely that this trend to allocate more spectrum for mobile services will continue.

Another clear trend is for the FCC to give its licensees more flexibility in how they use the spectrum that has been assigned to them. This increased flexibility includes the ability to provide different services and to use different technologies. As discussed before, the providers of trunked radio dispatch services have been given the right to offer interconnected mobile telephone service. Similarly, FM broadcasters have been given the right to offer radio paging services, utilizing unused spectrum within the channels that they have been assigned and, more recently, cellular and PCS providers have been given permission to not only provide mobile services, but fixed (e.g., wireless local loop) services as well.

These are all examples of increasing service flexibility. In terms of technology flexibility, there is also a clear trend to allow licensees to choose whatever technology they want as long as they don't cause harmful interference to other licensees. Thus, for example, the new PCS licensees are free to choose the multiple access technique that they think will best meet their business plans. This trend toward more flexibility stems from two thoughts. First, there is a feeling that the markets and technologies are changing so fast that the private sector is in a better position to choose what services to offer and what technology to deploy. Second, in terms of new spectrum to be auctioned, the value of the resource (and hence the revenues collected for the Federal treasury) is increased if the licensees have the ability to offer the most financially valuable service and the most cost-effective technology. For these reasons, many observers expect that such flexibility remain a key component of spectrum management policy in the future.

Still another clear trend is for the FCC to support spectrum allocations for unlicensed use. In the unlicensed radio bands, interference is controlled by regulating the technical characteristics of the equipment (e.g., by limiting transmitter powers to very low levels). Because the costs of licensing are eliminated and the other entry barriers are so low, manufacturers have produced a plethora of different equipments, including the extremely successful cordless telephones discussed earlier. Indeed, the small radios that are already embedded in such devices as crib monitors, garage door openers, intrusion alarms, and some wireless LANS operate as unlicensed apparatus under the Commission's rules. The FCC currently is considering proposals for a high performance, unlicensed service that would facilitate limited range but high-speed access to the Internet (or, more broadly, the National Information Infrastructure — NII). While the past success of the unlicensed services suggests continued support for them in the future, there are offsetting pressures. First, spectrum that is set aside for unlicensed use cannot be auctioned since no exclusive use is involved. Second, there is concern that, without the property-like rights that are associated with many of the licensed services, the licensees will not have the incentive to use the spectrum efficiently. Thus, while the trend toward support for unlicensed services seems clear, it must be moderated by these other considerations. Several years ago, the FCC received temporary authority to use auctions in assigning radio licenses. Thus far, the auctions have been regarded as very successful in terms of both assigning the spectrum quickly and raising money for the Federal treasury. For these reasons, plus the continued Federal budgetary pressures, many observers expect the auction authority to be renewed.

For several decades the FCC has strived to introduce more competition into the telecommunications industry generally and into the mobile radio services specifically. The Commission did so under legislation (the Communications Act of 1934) that was badly outdated and which did not contain language that called for reliance on competition rather than monopoly in the provision of telecommunications services. In early 1996, of course, the Congress passed and the President signed into law, the Telecommunications Act of 1996. Among other things, the new legislation mandated the promotion of competition and consumer choice in telecommunications services. Thus, it is clear that increased competition will continue to be the trend in the provision of mobile services as in telecommunications services more widely.

FUTURE TECHNOLOGICAL DEVELOPMENTS

The purpose of this section of the report is to speculate (hopefully in an informed way) on the longer term trends in wireless telecommunications technologies and services and to evaluate those trends in terms of their implications for people with disabilities. While it is very risky to speculate on future developments in a field that is changing as fast as wireless communications, there does seem to be a certain amount of concensus about broad trends.

Communicating Anytime, Anywhere, and in Any Mode

There seems to be a wide acceptance of the notion of being able to communicate anywhere, anytime and in any mode. This notion, or vision, of course, encompasses not only the wireline networks but, by necessity, wireless networks as well. Certainly this vision is consistent with technological, marketplace, standards-making, and policy/regulatory trends that are evident even today. For example, it is consistent with the medium term trends toward improved coverage, integrated service provision, and improved functionality as identified in Section IV.C.1. Realization of the vision will also be facilitated by the development of software programmable, multiband, multimode radios as described in Section IV.A.7. Software programmable radios will allow a subscriber unit to adapt its modulation, multiple-access method, and other characteristics to be able to communicate with a wide range of different systems and more efficiently support diverse voice, data, image and video requirements. Likewise, a single base station unit will be able to adapt its characteristics to support older generations of subscriber equipment while retaining the ability to support future developments through software changes. Thus, even current trends support the notion of the user being able to communicate anywhere, anytime and in any mode using the most efficient and economic means available.

Extending Multimedia and Broadband Services to Mobile Users

It has been widely observed that once people experience and get used to certain computer or communications services or capabilities in the office or at home, they soon want the same service or capabilities while they are away from those locations. This phenomenon is readily observable in personal computers where features and functions available on desktop systems quickly migrate to laptop computers that users can carry with them.

One clear trend in both computers and communications is toward supporting multimedia applications — i.e., applications that may combine sound, data (including text), image and video components. Thus, it is envisioned that as such multimedia applications grow in popularity, consumers will want similar multimedia capabilities delivered to them when they are away from their office or home.

Providing for the electronic transport of multimedia services is a challenging one because of the inherent differences in voice, data, image and video traffic as explained earlier. In the wireline network, a clear trend is towards the use of the asynchronous transfer mode (ATM) as a flexible format for handling, on the one hand, a mix of realtime, interactive, delay-sensitive traffic such as two-way voice and, on the other hand, non-realtime, less interactive, nondelay-sensitive traffic such as electronic mail. Multimedia traffic is also characterized by both continuous bit rate sources and variable bit rate sources. Examples of continuous bit rate sources include low-speed digital voice and high-speed digital video. Examples of variable bit rate or "bursty" sources include data files and digital images downloading wherein the need for capacity is not continuous or constant but, rather, varies from zero, when no information is requested, to very high, when the download of a file or image has been requested from the source.

ATM handles this mixture of continuous bit rate, variable bit rate, realtime, delay-sensitive, and non-realtime, delay-insensitive traffic by converting all of the information into a common format consisting of a sequence of fixed length cells. In other words, all of the traffic, regardless of type, is "chopped up" into short cells that are individually processed (switched). Each cell is 53-bytes long. The cell, in turn, consists of a header, which is five bytes long, and the information "payload" which is 48-bytes long. The short, fixed length cells can be switched at very high speeds using the routing information contained in the header. This high-speed switching, coupled with a call set-up process that limits the traffic intensity on the links between switches, means that end-to-end delays can be limited to acceptable levels. In short, ATM attempts to handle the different types of traffic with their different requirements based on (a) packetizing the information into short, fixed length cells, (b) routing or switching the cells at very high speeds, and (c) a call set-up process that assures against excessive congestion (and resulting delays) on the links that comprise the route established during that process. Because of its ability to handle different types of traffic, ATM is well suited for handling multimedia communications.

One of the advantages of ATM is that, just as in ordinary packet switching (and unlike in circuit switching), when a user is not sending information, he/she is not consuming network transmission resources. Moreover, when a user needs more capacity or bandwidth, he or she can acquire it simply by utilizing more cells (up to the limit imposed by the delay limitation on the shared links). This is often referred to as "bandwidth on demand" because the exact amount of bandwidth needed — no more, no less — can be allocated on a moment-by-moment basis.

Extending multimedia and bandwidth on demand capabilities into the wireless environment presents a number of challenges. For example, ATM in the wired network has been based upon the availability of broadband, high-quality, low error rate transmission links (e.g., fiber optic cable based links) and the assumption that the end terminals remain connected to a fixed port on the network. In the mobile, wireless world, available bandwidths (and, hence, transmission rates) are typically limited by spectrum scarcity and, as discussed at length in Section II.A., the transmission links between the mobile and base station are often characterized by high error rates and rapidly changing performance. In addition, a mobile terminal that is connected to one port on the network one minute may be connected to another one the next. Despite these challenges, there is a considerable amount of research interest in wireless ATM and an entire issue of IEEE Personal Communications was devoted to the topic. Given this interest, the technological and regulatory developments that have already been described, and user demands to extend fixed network capabilities into the mobile environment, it seems almost inevitable that the vision of delivering multimedia and broadband services to mobile terminals will be realized.

In a recent article describing European Union supported R & D projects involving the development of future generations of mobile communications concepts, systems, and networks, the writer summarized the above two long-term trends as follows:

From the user's perspective, the [program] will strive to ensure that current mobile services are extended to include multimedia and broadband services, that access to services are made without regard to the underlying networks, and that convenient, lightweight, compact, and power-efficient terminals adapt automatically to whatever air-interface parameters are appropriate to the user's location and desired services.

Embedded Radios

There is also a long-term trend toward, for want of a better term, embedded radios. This trend is comparable to the trend in microelectronics that is resulting in computer chips being installed in such ordinary items as automobiles, washing machines, sewing machines and other consumer appliances, and office equipment. They are even being installed in humans in the form of pacemakers and heart monitors. These devices are less visible than personal computers, computer workstations, and mainframe computers, but their impact on everyday life is significant nevertheless.

There is a less developed but comparable trend in radio devices as well. For example, small radio transmitters are being placed in key chains, and the transmitter is used to communicate with a small receiver in a vehicle so that its doors can be unlocked remotely. Also, small Global Positioning System receivers are being placed in other devices such as laptop computers so that the location of the device can be pinpointed. Similarly, small two-way radio devices are being placed in laptop computers or Personal Digital Assistants (PDAs) to enable them to communicate via the two-way paging systems or cellular-based CDPD systems described earlier. It has even been reported that vending machine operators are considering putting microprocessors and small two-way radio devices in their machines so that they can automatically report when they are out-of-service, are out of a particular product, have a full coinbox, or are being physically abused. This means that, in the long term, radio communications devices will increasingly be placed or embedded in other equipment just as computer chips have been placed in such ordinary items as consumer appliances. This will bring not only computer power to such equipment, but communications power as well.

Evaluation of Long-Term Trends

Like the medium term trends evaluated earlier, these long-term trends in wireless telecommunications technology and services would appear to generally bode well for Americans with disabilities. Certainly the ability to communicate anytime, anywhere, and in any mode has the potential of providing substantial benefits to disabled Americans. The long-term trend toward embedded radios also has the potential for providing substantial benefits because it will allow communications capabilities to be extended to not only general purpose devices utilized by the disabled, but also to specialized devices as well.

In a recent paper entitled "Access to the NII and Emerging Information Technologies by People with Disabilities," the authors described combinations of new technological developments that would provide powerful new capabilities and opportunities for individuals with disabilities. One potential example, a listening pen, was described as follows:

Listening Pen

An individual who is deaf may carry a small directional microphone which looks like a pen or is worn as part of their eyeglasses. When talking with someone, they would point the "pen" toward the individual's mouth. The speech would be picked up and sent out digitally over the net to powerful filter and speech recognition software running on a large computer. The result could be sent back to a small virtual display mounted to the deaf individual's glasses, which would project the image so that it appeared to float in front of them. In this fashion, the person who was speaking would have their words literally "written all over their face." Using voice print technology, it would even be possible for the speaker to be identified, if for example the person was sitting at a meeting where different people around the room were speaking in a mixed fashion. By using remote computing connected via wireless networks, the individual who was deaf could have access to much more powerful speech recognition algorithms than they would be able to or care to carry with them all day. In fact they may be paying a small service charge to use search recognition algorithms, owned and maintained by a network service bureau, which would otherwise be too expensive for the individual to afford and too rapidly changing for the individual to keep up with. The concept of the embedded radio would allow a small two-way radio to be built into the pen. It would allow (a) the speech to be sent to a powerful workstation computer containing powerful filter and speech recognition software and (b) the speech that has been converted to text to be received back from the workstation for projected display.

Communications in both directions, from the pen to the workstation and back to the display, would utilize an in-house wireless LAN. Or, if more mobilility were required, a larger, more powerful radio could be used with terrestrial and satellite based networks as described earlier. Finally, if it was a software defined or programmable radio, the end user transmitting and receiving equipment could be used for text telephone (TTY) service, to retrieve fax messages or voice mail messages that the network has converted to text, or to browse the Internet. In short, this is an example of where wireless technology can be used to extend network-based computer power to the end user "anytime, anyplace, and in any mode."

The same paper also described a hypothetical device called "The Companion." As the authors conceived it, this device containing computer-based artificial intelligence would assist people with cognitive impairments. This clever concept will not be described in detail here, but it is interesting to note that it includes communications capabilities. For example, the device would include a position location system to pinpoint the individual's location at anytime. By knowing the precise location of the individual (e.g., at home, at the bus stop, or at work), the device could tailor its assistance appropriately. With communications capabilities, the device could link to a more powerful central computer resource or to a human operator for "more serious problem-solving and for all of the situations where the limited artificial intelligence of the current device is not able to help."

The remaining one of the three long-term trends, the extension of mobile services to include multimedia and broadband services, would also appear to be beneficial to the disabled, with an important caveat. Like the population at large, the benefits of higher speed wireless (broadband) networks with shorter transmission delays are significant because they allow, for example, the downloading of more detailed information in a shorter time. For example, they would allow the downloading of a more detailed, color map to a lost motorist or to The Companion device carried by a person with a cognitive impairment.

The danger, and associated caveat, stems from the fact that the higher transmission speeds encourage more use of bandwidth intensive visual information which may make a service less accessible to visually-impaired individuals. And, more generally, as noted in a recent report by the Blue Ribbon Panel on National Telecommunications Policy, "Graphical User

Interfaces (GUIs) and touch screens, which are becoming so prevalent, cannot be used by blind, visually-impaired and some mobility impaired individuals." Similarly, the addition of voice messages into a multimedia computer-based presentation may create barriers for deaf or hard of hearing people while the addition of video clips may create barriers to visually-impaired people.

Thus the extension of mobile services to include multimedia and broadband services is a classic example of a two-edged sword. On the one hand, they hold the promise of significantly increasing the ability of people with some types of disabilities to access and use information more readily and to communicate more widely with all members of society. Properly applied, the technology can make their disabilities invisible or irrelevant and thereby allow them to more fully participate in — and contribute their talents to — the community at large. On the other hand, they can prevent persons with certain types of disabilities from gaining access to needed information, from using the information, and from communicating more readily. They can also improve the accessibility or usability for individuals with certain disabilities while lessening accessibility and usability for individuals with other disabilities. It should be emphasized that these advantages and disadvantages of these broadband and multimedia capabilities for the disabled are not unique to wireless. They apply equally well to wired services. Viewed in that way, wireless becomes almost totally an advantage for the disabled as it is for the public at large.

4

Tools and Technologies for Mobile Learning

M-LEARNING

The term m-learning or "mobile learning", has different meanings for different communities, that refer to a subset of e-learning, educational technology and distance education, that focuses on learning across contexts and learning with mobile devices. One definition of mobile learning is, "any sort of learning that happens when the learner is not at a fixed, predetermined location, or learning that happens when the learner takes advantage of the learning opportunities offered by mobile technologies." In other words, with the use of mobile devices, learners can learn anywhere and at any time.

M-learning technologies include handheld computers, MP3 players, notebooks, mobile phones and tablets. M-learning focuses on the mobility of the learner, interacting with portable technologies, and learning that reflects a focus on how society and its institutions can accommodate and support an increasingly mobile population. There is also a new direction in m-learning that gives the instructor more mobility and includes creation of on the spot and in the field learning material that predominately uses smartphonewith special software such as AHG Cloud Note. Using mobile tools for creating learning aides and materials becomes an important part of informal learning.

M-learning is convenient in that it is accessible from virtually anywhere. M-learning, like other forms of e-learning, is also collaborative. Sharing is almost instantaneous among everyone using the same content, which leads to the reception of instant feedback and tips. This highly active process has proven to increase exam scores from the fiftieth to the seventieth percentile, and cut the dropout rate in technical fields by 22 per cent. M-learning also brings strong portability by replacing books and notes with small RAMs, filled with tailored learning contents. In addition, it is simple to utilize mobile learning for a more effective and entertaining experience.

Mobile learning, or "m-learning", offers modern ways to support learning process through mobile devices, such as handheld and tablet computers, MP3 players, smartphones and mobile phones. It presents unique attributes compared to conventional e-learning: personal, portable, collaborative, interactive, contextual and situated, it emphasizes "just-in-time-learning" as instruction can be delivered anywhere and at anytime through it. Moreover, it is an aid to formal and informal learning and thus holds enormous potential to transform the delivery of education and training.

Mobile learning is emerging as one of the solutions to the challenges faced by education. UNESCO's programme of activities is therefore based on growing partnerships geared towards exploring how mobile technologies can enable the achievement of Education For All. Its partners include Nokia and the U.S. Department of State.

M-LEARNING: SMALL, ENGAGING AND AT YOUR LEISURE

m-learning is mobile learning. It is about using the massive growth of mobile technologies to benefit learning and learners.

There are several exciting projects underway around the globe at the moment, focusing on different perspectives of m-learning. They cover a broad range of technologies and learners and are starting to show some exciting results.

Here are some scenarios for m-learning:

- I am bored and waiting for a bus. I am taking my driving theory test in a week and am a bit nervous about it, so I load a game onto my phone that lets me practise 'quiz'-style questions while I wait.
- I see a poster on a wall challenging me to improve my maths. I try the 15 questions, text my answers to the number at the bottom and get an instant assessment.
- I am studying a foreign language. To practise listening, I call up a phone number that simulates various dialogues with me (e.g. buying a bus ticket) and gives me feedback on my under standing or even on my pronunciation!

Some others using slightly better phones:

- My class of students does a mobile project. I send them around town with a 'treasure map' and as they solve the various puzzles I have left them they take photos of the solution and text them to me. All the pictures and texts are automatically collected onto a web site, so that when they return they can show their friends what they did.
- I am dropping out of school. I take an assessment and it seems I need to improve my reading, but I am very unmotivated. As part of a technology trial I am lent a PDA-phone with a collection of learning materials matched to my needs, but designed for the small screen: bright, brief, light-hearted, loud. I get to keep the PDA-phone for a couple of weeks, in the process learning to write and draw on it, as well as surfing the web and taking photos!

In fact, there are a huge number of possibilities for these PDA-phones, such as sharing pictures or documents, collaborative game playing, reading and writing. All of these examples are real. We have got them working as part of the m-learning project. We are using the seductive power of these new technologies to re-inspire young learners who are dropping out of traditional learning. Research and development has been ongoing for the last two years and many learners have already been trying out these approaches and contributing to their development.

WHAT HAVE WE LEARNT?

Technology balance: m-learning covers a wide range of technologies. New phones and PDAs come out every few weeks and each has more power than its predecessor. Finding a balance between focusing on a minority (but powerful) phone or a generic (but very outdated) technology standard is essential. That balance will differ depending on your project, but being clear about it from the start will help avoid problems later.

Not a PC: It is easy to assume that the learning process will be similar to that using a PC, only at a smaller size. This is not true. In fact both the limits of the technology and the lack of control you have over how and when the learning happens means that you need a different learning model.

Bite sized: If you learn on the move, you do it in short sharp bursts and you are often surrounded by distractions (some are on your phone, like the web or other games) so learning must be engaging.

Keep a spread: Developing materials that can be used in different forms across different technologies will give the best mileage and stop them getting out of date when new phones are released.

Blended learning: The best learning materials are short. They work best when part of a bigger solution – like with the student quoted above waiting for the bus: the phone didn't teach him all the theory, it just supported him as he tried to remember it.

Collaboration: In every single trial, the learners engaged most with learning that they could do together, either by sharing phones or by passing things between phones. Try to build learning around this.

What next?

Currently, 100 gadgets of various shapes and sizes are being sent around Britain, Italy and Sweden to (we hope) inspire young learners in all of those places. We have developed various tools and engines to make it quicker to add more materials and have already been contacted by a number of organisations who are interested in running their own m-learning projects.

Judging from the surge in interest we have seen in the past six months, it feels as if m-learning is starting to move into mainstream delivery. Initially we were talking to researchers, but now we are being approached by large government departments and organisations interested in real training.

What else?

We have had some very interesting discussions with deaf learners who are very advanced users of text messaging and are hungry for learning opportunities. We haven't yet touched on another rapidly growing aspect of m-learning: 'location aware' technology. We did not use it in our trials because of the wide range of locations we are working in, but it is a powerful way to find out facts and is already in use. For example, as I walk around the museum, my phone knows where I am and can tell me facts about the exhibit I am looking at.

Moving to the Mainstream

The commercial potential of m-learning is becoming apparent in the UK. The current budget for post-16 education and training (below university level) is around £9 billion. A significant amount of this is targeted at what are called 'hard-to-reach learners'. These include the young disaffected people who were our project's original target audience, as well as people in low level jobs; in highly mobile jobs with unpredictable hours that make it hard to commit to a fixed programme; and also people who have limited access to PCs and the Internet. m-learning offers the opportunity to reach out to these people wherever they are and do things that are useful and meaningful. Funding follows outcomes and we have demonstrated some powerful outcomes using these technologies.

Current examples include:

- *Health and safety in the workplace:* The largest topic that is the subject of training in the UK. We are now starting fully funded projects which allow people to practise the knowledge tests that are legally required of employers, using mobile technology to deliver the questions.
- *M-learning for learning disability:* A major UK project to develop learning services for people with various disabilities is underway; there is great interest and excitement from people providing services to the visually impaired, the hearing-impaired, those with major physical needs (such as wheelchair and switch users) and many others. This project is developing mainstream content for such groups as part of its official work programme.
- *M-learning for health:* On top of the UK's mainstream training budget there is also a separate set of initiatives that are about developing a more healthy society; these are often associated with regeneration or with preventive medicine. The potential of mobile devices here is being explored in a number of projects; one that we are involved with is using a mix of magazine-style health information with phone-based quizzes and assessment to reach pregnant teenagers in inner-city Birmingham.

- *M-learning for health professionals:* We are beginning work with a university which is keen to apply mobile technology to the training of health professionals, using some of the more advanced handsets that are capable of creating and sending video clips as well as running sophisticated software programmes. On-demand information, the ability to share visual images and knowledge testing are just three examples of what is being explored.
- *M-learning for work-based learning:* The government has a major commitment to promoting learning in the workplace, where there is often very little technology that can support conventional e-learning. PDAs and more sophisticated handsets with multimedia capabilities are a much more satisfactory solution than having a fixed workstation; there is considerable potential in providing learners with the PDAs or handsets with relevant content already loaded for them to use within a structured, blended learning process that includes face-to-face, phone and text-based and online tutor support.

It's important to stress that these opportunities are representative of the mainstream, not small isolated examples. All are carefully structured to comply with mainstream funding formats and requirements. We are in discussion with several of the bodies responsible for vocational education in specific sectors (such as distribution, retail, construction and care) and are developing blended solutions with them that have a mobile element. In our view, this is the early phase of a very large and growing market: blended learning needs to adopt a mobile component.

What is important about this project is that it has a focus on presenting learning on mobile phones and not on other wireless devices. This is because of its social context. This context is presented as addressing social and educational problems in young adults. The m-learning project addresses three social/educational problems relating to many young adults in the EU:

- Poor literacy/numeracy - see e.g. Improving Literacy and Numeracy: A Fresh Start
- Non participation in conventional education/training
- Lack of access creating ICT "haves"/"have nots" resulting in inequality of opportunity

The project will develop prototype products to provide information and modules of learning via inexpensive portable technologies which are already owned by, or readily accessible for, the majority of EU young adults. This primary social focus is on unemployed and unemployable British youths who refuse to attend colleges or training centres. None of these youths have lap top computers or PDAs. All have mobile phones, which they use constantly.

ADVANTAGES OF MOBILE LEARNING

No other learning approach matches the integrated, continuous flow of m-learning: *Seamless access to learning resources*: With mobile learning, you can learn and study anywhere - from the classroom to your desktop, or laptop to your pocket. A true mobile learning system allows users to take a course on any device.

Freedom, power, and choice: M-learning students can choose where, when, and how they'll study. The new range of options includes online synchronized, online self-paced, downloaded courseware, and computer-based training. M-learning offers new levels of freedom with the ability to exercise control over learning patterns.

Organized productivity: With only a cell phone, handheld device, PDA, or hybrid unit, users can access administrative functions, download courses, and review their learning history through a learning management system. M-learning offers an efficient way for learners to access key information and maximize their time.

Flexible, portable convenience: The ability to customize learning schedules is a key advantage to m-learning. Learners are not restricted to a specific physical environment, a particular delivery channel, or a fixed set of times for undertaking training and education. Using the latest technology, students can update their knowledge base on a just-in-time basis to prepare for meetings or presentations.

INTRODUCTION TO MOBILE LEARNING (M LEARNING)

Describes the mobile technology available today, compares the devices currently on offer, lists the benefits that the technology could bring to education, explains the accessibility issues, lists some ongoing projects, and provides links for further information.

The term mobile learning (m-learning) refers to the use of mobile and handheld IT devices, such as PDAs, mobile phones, laptops and tablet PCs, in teaching and learning. As computers and the internet become essential educational tools, the technologies become more portable, affordable, effective and easy to use. This provides many opportunities for widening participation in and access to ILT, and in particular the internet. Mobile devices such as phones and PDAs are much more reasonably priced than desktops, and therefore represent a less expensive method of accessing the internet (though the cost of connection can be higher), and the introduction of tablet PCs now allow mobile internet access with equal, if not more, functionality than desktops.

Benefits of mobile devices for education

Generally, PDAs offer greater functionality than mobile phones and similar advantages to tablet PCs, though tablet PCs are more robust than PDAs and offer additional features. PDAs and tablets, and to some extent mobiles, can be used in many educational settings and accomplish many different educational tasks. Most mobile devices are useful in education both as administration, organisation and teaching aids for practitioners, and also as learning support tools for students.

Students can interact with each other and the practitioner instead of hiding behind large monitors. Much easier to accommodate several mobile devices in a classroom than several desktops, as require far less space. PDAs or tablets holding notes and e-books are lighter, less bulky and easier to carry than bags full of files, paper and textbooks, or even laptops. The handwriting recognition software in PDAs and tablets have been found to help improve students' handwriting skills. Handwriting with the stylus pen is more natural than using keyboard and mouse. Can draw diagrams, maps, sketches directly onto a tablet, using standard software

Can take notes directly into the device during outdoor lessons or on field trips, either typed, handwritten or voice. Electronic registration and inputting data in practical lessons or outdoors where desktops are not appropriate or too cumbersome eg. science experiments, kitchens, farms. Shared assignments and collaborative working, so several students and the practitioner can pass the device around a group, or "beam" the work to each other using the infrared function of a PDA, or a wireless network such as Bluetooth.

Practitioners can more easily and naturally annotate work using the pen. Can be used anywhere, anytime, including at home, on the train, in hotels - such places are conducive to learning because you cannot be disturbed by meetings, you are often alone, it might be quiet - this is invaluable for work-based training. Stylus pens are much more natural for web browsing - click directly on links etc with the pen instead of using a mouse. Can trace an image directly onto the tablet's screen. Engaging learners - young people who may have lost interest in education do like mobile phones, gadgets and games devices such as Gameboys. Increases motivation and personal commitment to learning if a student can "own" a device and take it with him/her wherever he/she goes, and encourages responsibility. May contribute to combating the digital divide, as generally cheaper than desktops, especially mobiles and PDAs.

Just-in-time learning/reference tool for quick access to data in the field eg. accessing step-by-step guides to help you achieve a task. SMS can be used to get information (eg. timetable changes) to staff and learners more easily and quickly than phone calls or email, for example. As assistive technology for learners with learning difficulties and/or disabilities.

Disadvantages

Small screens limit the amount and type of information that can be displayed (mobiles and PDAs). Limited storage capacities, especially mobiles and PDAs. Batteries require regular charging, and data can be lost on some devices if this is not done correctly. Lack of common platform (eg. Different sized screens - horizontal screens with some handheld computers, small square screens with mobile phones etc), so difficult to develop content that will work anywhere. More easily lost or stolen than desktops, more attractive to thieves than paper notes. Much less robust than desktops.

Difficult to use moving graphics, especially with mobile phones, although 3G and 4G will eventually allow this. More difficult to upgrade and limited potential for expansion with some PDAs. Fast-moving market so devices can become out of date very quickly

Lack of connectivity and interoperability, though new technologies such as Bluetooth are beginning to address this issue Security issues when accessing wireless networks via mobile devices. Bandwidth may degrade with a larger number of users when using wireless networks. Practitioners may require extra training in order to be able to use the devices effectively. Difficulties with printing, unless connected to a network

N Deviney and C Von Koschembahr, Learning goes mobile

Deviney and von Koschembahr begin their presentation with some applications for the pharmaceutical industry:

Imagine a pharmaceutical sales representative preparing to meet with a client.

While he waits for his meeting to start, he uses his personal digital assistant for communications and e-learning. With the information accessible to him, not only does he stay up-to-date on the market issues essential to his position, but he also receives regular notification from message boards, news portals and his employer.

The PDA buzzes in his pocket. This time, the notification tells him his competitor just released a new pain medicine that will compete with the drug he is selling. He accesses the press release and news coverage and reviews key differentiators between the products that his company posted for his review. Based on a personalized profile - created by him - this sales representative learns while he waits for a client, on his lunch break and during down time. He can receive "just in time" information, such as a notification the moment his company gets governmental approval for one of the new drugs he is trying to sell. His favorite news organizations also notify him when breaking news happens. His employer regularly updates material on the competitive landscape for the sales staff.

This is the future of learning. It demonstrates the role that mobile learning can play for both organizations and individuals using technology that many Americans already carry with them today. The situation outlined above is based on a capability called "profiled notification." We receive profiled information everyday from Yahoo, CNN and other portals that allow us to pick and choose what information we want to receive on a regular basis, and have the information delivered to our e-mail boxes as it becomes available.

Profiled notification, as used by the pharmaceutical sales representative, takes information delivery to the next level by teaching and informing on a more urgent basis. He is already proactively enrolled in learning events and information sharing portals, but without profiled notification, he does not know when his learning events have changed. As a mobile employee, profiled notification allows him to stay in touch with information most valuable to him. By notifying him via his mobile device, we reach the sales representative with the information he wants, when and where he wants it. For many professionals - whether in sales, retail, banking, government or health care - having information on demand, at their fingertips, changes the way they work.

Deviney and von Koschembahr continue their presentation with an overview of the market:

Many employees are equipped. More than 150 million Americans carry a mobile phone. According to IDC, that number will grow to more than 180 million by 2007. Four per cent of Americans are trading in their landline phones in for cost-effective wireless deals. And many carry smart phones that are alternatives to hand-held computers and PDAs. Overall, mobile devices are coming equipped with more advanced features - such as streaming video, color-touch screens, Internet browsers and compatibility with desktop applications - that make profiled notification and mobile learning not only possible, but practical.

Given such advances, Stamford, Conn.-based research firm Gartner reported in November 2003 that enterprises will increase the use of mobile applications by 80 per cent during 2004. Market trends indicate a significant number of employees are already using their mobile devices to access e-mail, search the Web, organize calendars, read the news or access documents during down time. Although manufacturers are building PDAs, cell phones and hand-held computers that deliver applications and content, the burden lies on employers to deliver the information and materials that keep employees engaged and competitive. Companies have the opportunity to rethink how employees can use their hand-held devices to enhance productivity.

They give their view of the role of mobile learning:

The intention with mobile learning is not to take the place of the classroom or hands-on experience, but to enhance its value. Mobile learning offers another way to deliver content and to embed learning into everyday work flow. Employers can develop learning materials once and deliver it in many ways, such as creating small, consumable bytes of content that can be delivered to their traveling employees or sales teams. Companies can improve the quality and productivity of the time spent in the classroom by offering material for advanced review; or they can reach mobile employees through interactive Web lectures; or they can push "just-in-time" information to enable staff to stay competitive. In the pharmaceutical scenario, the information is urgent and critical for the sales team. However, in other cases, mobile learning is also a way to keep employees engaged. For example, an employee waiting for 20 minutes to catch a flight or a train might access her company's best practices database. She learns how her co-worker grew a small, regional account into a multinational million-dollar client. She decides to request a meeting with her colleague to brainstorm ways she may grow her latest customer win. Even though the information was not urgent, it was delivered in smaller portions to a device this employee had in her briefcase or purse, and she was able to stay engaged and relate the material directly to her work activity.

Consider this: A significant number of workers with poor training opportunities look for a new job within one year. In retail, training new staff can cost several thousand dollars, and having well-trained and satisfied employees is critical to the profitability of an organization. In the electronics retail industry turnover is a huge challenge. Many former electronics retail employees complain they felt ill-prepared to help customers or answer questions on the retail floor. To address these training issues, a major electronics retailer may opt to use a mobile learning program to train new sales associates.

Traditionally, new employees spend several hours in a back office reading a binder or accessing a learning portal to study various products. The products are not in front of the employee, and for electronics retail associates, there are often hundreds of products to understand. This retailer takes a different approach and equips its associates with a hand-held PDA and bar-code scanner. Employees start and finish their training on the sales floor. During down time, they find a product, scan a bar code and take a five-to-10-minute training module in front of the product. Or, when servicing a customer, the associate can do "ad hoc" learning.

Transitioning to Mobile Learning. Imagine the implications of mobile learning for other industries. For the military, it means soldiers receive mission-critical information and briefings in the field. For a finance manager, it means unobtrusively providing mortgage rate updates during a client meeting. For a new bank manager, it means coaching simulations and management training are available anytime, in advance of in-person meetings with employees. The bottom line is that mobile learning offers organizations the ability to keep their employees engaged in ongoing learning activities that will enhance their productivity and effectiveness while they work. It enables client-facing teams to be better informed and more responsive to client needs, giving them a competitive edge.

Organizations need to align their targeted mobile learning initiatives to support their business priorities, as well as understand job roles and employee needs. With the right foundation in place, a new world of delivery and communication will empower organizations to create a competitive workforce for the 21st century.

DEFINITION OF MOBILE LEARNING

In providing a definition of mobile learning one is faced with tensions between functionality and mobility. The technologies involved in e-learning and m-learning (computers, laptop computers, PDAs (Personal Digital Assistants)/handhelds/ palmtops, smartphones and mobile phones can be arranged on a continuum:

FUNCTIONALITY		MOBILITY		
Computers	Laptop Computers	PDA'S Handhelds Palmtops	Smartphones	Mobile Phones
<—— E-LEARNING<——>		<—— M-LEARNING<——>		

Fig. Relationship of m-learning to e-learning

Many experts on mobile learning, especially in the United States of America, include laptop computers in their definition of mobile learning, but I disagree. I consider that it is the degree of mobility that is the defining element in mobile learning, even if one has to sacrifice functionality. Therefore, the term mobile learning should be limited to education and training on devices one can comfortably carry around, in one's hand or in one's pocket, and therefore I do not think that the definition of mobile learning should be extended to include laptop computers.

FIVE EXAMPLES

I would like to share with you at this stage five examples of mobile learning in action. four of these are European Commission-funded projects and one is a development in Africa.

The five projects of mobile learning in action are:

1. The *From e-learning to m-learning* project led by Ericsson, Ireland
2. The *Mobile learning: the next generation of learning* project led by Ericsson, Ireland
3. The *M-learning project* led by the United Kingdom LSDA (Learning and Skills Development Agency)
4. The *MOBILearn* project led by Giunti Ricerca of Genoa, Italy
5. The *M-learning in rural Africa* project of the University of Pretoria, South Africa.

M-LEARNING DEVELOPMENTS

M-learning is a natural extension of e-learning and has the potential to make learning even more widely available and accessible than we are used to in existing e-learning environments. The role that communication and interaction plays in the learning process is a critical success factor. It is within this context that m-learning can contribute to the quality of education. It offers opportunities for the optimization of interaction between lecturers and learners, among learners and among members of COPs (communities of practice).

Wireless and mobile technologies also make it possible to provide learning opportunities to learners that are either without infrastructure for access (example rural learners) or continually on the move (example business professionals).

Why m-learning in Africa?

One's first impressions and perceptions when thinking about the ideal target market for m-learning would probably look like this: *a first world learner population that is already highly ICT literate, uses the latest handheld device and is either in full-time employment or merely prefers studying at their own pace, place and time.*

This description does not fit the majority of learners in Africa though. Why then m-learning in Africa? Well, the answer is quite interesting. Because of the lack of infrastructure for ICT (cabling for Internet and telecom) in certain areas in Africa, the growth of wireless infrastructure is enormous - even more rapid than in many first world countries.

The East African (July 8, 2002) reported that: '...the communications sector in Uganda is growing rapidly. Nua Internet Surveys (July 15, 2002) reported that, according to the National Information and Communication Technology Policy, the number of mobile phone subscribers in Uganda grew from 3,500 in 1996 to a total of 360,000 in 2002.'

Wachira (2003) reported the following about Kenya: 'When Vodafone UK sent Michael Joseph to Kenya in July 2000 to set up Safaricom, a cell-phone service operator jointly owned by Telkom Kenya, he did not expect the subscriber base to grow beyond 50,000 connections. Today, both Safaricom and rival KenCell Communications (partly owned by Vivendi) have nearly 1.3 million cell-phone subscribers. This set-up is deeply rooted in the traditional African communal mode of living, which many urban dwellers haven't abandoned.'

Shapshak (2002) reported that the adoption rate of mobile technologies in Africa's developing countries is among the highest rates globally and forecasts estimate almost 100 million mobile users in Africa by 2005. Between 1997 and 2001, the number of mobile phone subscribers in Africa annually had a triple-digit growth rate.

Mobile subscribers in Africa have risen further and increased by over 1000 per cent between 1998 and 2003 to reach 51.8m (ITU, 2004). It is thus obvious that the adoption rate of mobile technologies is exceptional in Africa.

According to Brown (2004), we can therefore differentiate between two ideal target markets for m-learning: learners that are either without infrastructure and access or learners that are continually on the move. In other words:

- First world learners who are the workforce on the move with state of the art mobile devices and
- Third world rural or remote area learners who have mobile phones.

THE BASIS FOR MOBILE LEARNING

The justification of mobile learning comes from the 'law' of distance education research which states that 'It is not technologies with inherent pedagogical qualities that are successful in distance education, but technologies that are generally available to citizens'.

A typical example is the 12" laser discs of the early 1990s. These laser discs had excellent pedagogical possibilities and excellent courses were developed for them especially in the field of ESL (English as a Second Language), but they were not successful because not enough people owned one. Never in the history of the use of technology in education has there been a technology that was as available to citizens as mobile telephony. The statistics are stunning:

- Ericsson and Nokia tell us there are 1.500.000.000 of them in the world today for a world population of just over 6 billion. Sales will continue as ownership of the latest model becomes of social importance and multi-ownership is already an important phenomenon. The roll-out of 3G will create pressure for the purchase of 3G compatible handsets.
- 50 per cent of all employees spend up to half of their time outside the office
- more than 525.000.000 web-enabled phones were shipped by 2003
- worldwide mobile commerce reached $200 billion by 2004
- multi-purpose handheld devices (PDAs and smartphones) will outsell laptop/desktop computers combined by 2005.

The urgency of the development of mobile learning is enhanced by the imminent arrival of 3G wireless technologies which have already been rolled out in certain EU countries and will shortly be available to all.

The arrival of 3G will bring:

- People will be able to manage better their time and personal work: on a train, at airports, while waiting etc
- Applications that run today on a computer will be able to run on a phone
- The Internet and the WWW will be accessible directly to citizens on their phones
- A wide range of applications will run on phones rather than on computers: electronic passport visas can be mailed directly to the phone, electronic payments can be made by phone not computer.

- 3G will guarantee video connections over the air, in real time as opposed to the fragile connections of today
- Citizens will be able to work from anywhere with their phones
- The data rates available from 3G will make large data transfers from phones practical
- 3G is for laptops and wireless LANs as well as telephony but is only available in hotspots today. For covering citizens in their homes with the data rates they need to support the services they want they must have 3G.
- 2G and 2.5G provides coverage not capacity. WiFi supplies capacity not coverage. 3G provides coverage and capacity.
- In 3G the bandwidth for applications is available to the phone so that the only limitation to applications is the imagination. It is important that learning and training do not miss out.

The humorous version of what 3G will bring to phones is Girls, Games and Gambling but it is clear that the advent of 3G will impact favourably on mobile learning. Advantages 3G will bring include: handhelds rather than desktops; decreasing costs; increase in battery life; increase in memory; constant always-on connectivity; increase in functionality; video; high definition TV; digital radio; software as for desktops; voice recognition; global tracking systems; size of devices; scientific calculation; recording of evidence and assessment; the future is wireless.

Recent research on audience characteristics published by the BBC in Britain shows the ubiquity of mobile devices especially in the 16-24 age group, the university student age group.

In the BBC research this group characterises the possession of a mobile phone as a 'necessity'. Thus it can be taken as a given that all students in all European further and higher education institutions possess one.

DEFINITION OF MOBILE LEARNING

In defining mobile learning one confronts tensions between functionality and mobility. There is a continuum form the point of view of functionality in the devices used for e-learning and m-learning. This continuum goes from desktop computers to laptop computers to PDAs or handhelds or palmtops to smartphones to mobile phones. There are many, especially in the United States of America, who include laptop computers in their definition of mobile learning.

I disagree. I feel that in the definition of *mobile* learning the focus should be on mobility. Mobile learning should be restricted to learning on devices which a lady can carry in her handbag or a gentleman can carry in his pocket.

I therefore define mobile learning as 'the provision of education and training on PDAs/palmtops/handhelds, smartphones and mobile phones.'

FUNCTIONALITY		MOBILITY		
Computers	Laptop Computers	PDA'S Handhelds Palmtops	Smartphones	Mobile Phones
<—— E-LEARNING<——>		<—— M-LEARNING<——>		

Fig. Functionality and mobility in a definition of mobile learning

One of the characteristics of mobile learning is that it uses devices which citizens are used to carrying everywhere with them, which they regard as friendly and personal devices, which are cheap and easy to use, which they use constantly in all walks of life and in a variety of different settings, except education.

MOBILE LEARNING PROJECTS

There are four major projects in mobile learning funded by the European Commission in Brussels. These are:

- The Leonardo da Vinci project *From e-learning to m-learning* led by Ericsson Education Dublin.

The *From e-learning to m-learning* project addressed the development of courseware for mobile phones, smartphones and PDAs. The main pedagogical concepts of developing mobile learning for PDAs were solved in the project *From e-learning to m-learning,* in which a comfortable didactic environment was created by using Microsoft Reader Works, providing each student with Microsoft Reader software to display the content and which was adjudged highly satisfactory by surveys of students who had studied a full course by mobile learning on a PDA. The full course of text and readings measured 1000 A4 pages and was easily held by the memory of a standard PDA like the HP Compaq iPaq 5000 series. The project also had success in developing courseware for phones and incorporated surveys of student satisfaction with mobile learning for students from Ireland, Norway, Germany and Italy.

- The Leonardo da Vinci project *Mobile learning: the next generation of learning* led by Ericsson Education Dublin.

The main activities to achieve the products of acceptable courseware for smartphones in the *Mobile learning: the next generation of learning project* are installing a web-authoring tool like Macromedia Dreamweaver MX Version 1.0, installing a desk-top browser e.g. Opera 6.31 that has page rendering characteristics similar to a mobile phone, using XHTML 1.0 Transitional to code the web pages, using Cascading Style Sheets (CSS) to separate presentation style from document content, investigating the level of support for Javascript on phone browsers, arranging each course page as a vertical column of content of at most 208 pixels, dealing with Table elements, designing the user interface so that the limited screen space available is utilised as efficiently as possible, designing concise neat diagrams to fit the small screen, using Adobe Photoshop to produce light weight GIFs for mobile devices. Progress was also made in the development of courseware for PDAs with one of the partners, NKI from Bekkestua, Norway announcing that by developing server-side code for their system, they had produced mobile learning versions of all 400 of their e-learning courses and were offering them to their mainline students.

- The IST project *M-Learning* led by the United Kingdom government Learning and Skills Development Agency (LSDA)

This project had an important social dimension. It recognised that there were in the United Kingdom many 16 to 20 year old youths who were unemployed and had urgent needs for additional training, but who refused to attend a training centre or college. They were unemployable and refused to attend training. They all had, however, a mobile phone which they used constantly. The project, therefore, set out to develop courses for them on their mobile phones in the fields of literacy, numeracy and social skills. The focus of the project was on mobile phones, as this type of student did not possess either smartphones or PDAs.

One of the organisers of the project writes: 'The commercial potential of m-learning is becoming apparent in the UK. The current budget for post-16 education and training (below university level) is around £9 billion. A significant amount of this is targeted at what are called 'hard-to-reach learners'. These include the young disaffected people who were our project's original target audience, as well as people in low level jobs; in highly mobile jobs with unpredictable hours that make it hard to commit to a fixed programme; and also people who have limited access to PCs and the Internet. m-learning offers the opportunity to reach out to these people wherever they are and do things that are useful and meaningful. Funding follows outcomes and we have demonstrated some powerful outcomes using these technologies. Current examples include: Health and safety in the workplace: the largest topic that is the subject of training in the UK. We are now starting fully funded projects which allow people to practise the knowledge tests that are legally required of employers, using mobile technology to deliver the questions. m-learning for learning disability: a major UK project to develop learning services for people with various disabilities is underway; there is great interest and excitement from people providing services to the visually impaired, the hearing-impaired, those with major physical needs (such as wheelchair and switch users) and many others. This project is developing mainstream content for such groups as part of its official work programme' (Stead 2003).

- The IST project *MOBILearn* led by Giunti Ricerca of Genoa, Italy.

This is a very large project led from Italy and counting a wide range of at least 20 European universities among its members. The objectives of this project are: The definition of theoretically-supported and empirically-validated models for: Effective learning/teaching/tutoring in a mobile environment;

Instructional design and eLearning content development for mobile learning.The development of a reference mobile learning architecture that is attractive to key actors in Europe and beyond, and that supports: Human interfaces adaptive to the mobile device in use and the nature (e.g. bandwidth, cost) of the ambient intelligence that is available in a given location; Context-awareness tools for exploiting context and capturing learning experience; Integration of mobile media delivery and learning content management systems; Collaborative learning applications for mobile environments. The development of a business model and associated implementation strategies for successful EU-wide deployment of mobile learning, starting from: A study of existing business models and market trends; An appraisal of the external environment. Large-scale use of project results by all interested parties in Europe

These projects were *projects.* That is, they were research undertakings to set out the first building blocks of a new sector of education and training provision. It is now time for mobile learning to emerge from its project status and enter into mainstream education and training. Excellent work has been done by the projects listed above but until mobile learning enters the mainstream it remains a fragile and research-based undertaking.

The trouble with *projects* is that they tend to collapse and disappear when the project funding is discontinued.

THE FAILURE OF MOBILE LEARNING

Why has mobile learning failed to emerge from its project status and failed to take its place in mainstream provision? Why does it remain at the research project level and not emerge into a serious form of provision?

The problem is that wireless applications are being developed for wireless devices for all walks of life. Learning and training do not figure in these developments. Learning and training do not seem to be high on the list of applications that are receiving attention today. It is essential for mobile learning that developments in education keep pace with developments in other fields.

As relayed by innovation experts, new ideas and inventions only become innovations when the ideas or inventions are adopted and utilised by the market. Perhaps one of the explanations for the development of m-learning lies in the fact that the role that communication and interaction plays in the learning process is a critical success factor in contemporary educational paradigms. M-learning appears to thrive within the contemporary constructivist and social constructivist paradigms because it enriches both synchronous and asynchronous communication and interaction.

Isn't it strange that all students enrolled in higher and further education institutions today have frequent needs for information from their institutions about timetable changes, assessment deadlines, feedback from tutors and other urgent administrative details? Nearly all of these students carry a sophisticated communications device which they use constantly in all walks of life except in their education or training programme.

Isn't it strange that all higher and further education institutions today have frequent needs for providing information to their students about timetable changes, assessment deadlines, feedback from tutors and other urgent administrative details? Nearly all of these students carry a sophisticated communications device which they use constantly in all walks of life except in their education or training programme.

The answer to these questions about the failure of mobile learning to move from project status into the mainstream is well known. It is that mobile learning is not seen as a satisfactory revenue stream for the telecommunications operators. The urgent need for mobile learning is to emerge from its fragile project status and convince the telecommunications operators that it represents a viable and valuable revenue stream.

A Matrix for Mainstream Provision

One can develop a nine-point matrix for the use of mobile learning in mainstream education and training. One axis is made up of the three types of devices that make up mobile learning provision:

- PDAs
- Smartphones
- Mobile phones.

The other axis is the types of education provision that can reasonably be provided by mobile learning:

- Mobile learning academic administration SMSs. This provides universities and other institutions with structures to send SMSs to all their students, or all students in a particular grouping about timetable changes, examination deadlines, assignment results, changes of procedure that all institutions need to immediately communicate to their students. This can also be used for students who need information from the institution in the form of FAQs, or answers contained in databanks.
- Mobile learning academic summaries. These are 4 to 5 screen academic summaries of content, examination hints, assessment questions for course revision, guidelines for particularly difficult parts of a course or counselling provision for students in need.
- Full modules by mobile learning. This is the presentation of full courses, or full modules of courses, on mobile devices.

This gives the following possibilities:

- Mobile learning for academic administration on PDAs. This has already been achieved but the number of university students possessing a PDA may be limited.

- Mobile learning for academic administration on smartphones. This has already been achieved but the number of university students possessing a smartphone may be limited.
- Mobile learning for academic administration on mobile phones. This has already been achieved particularly by the Faculty of Education at the University of Pretoria in South Africa. Extensive numbers of mobile learning for academic administration communications have been sent out to teachers in country schools doing post-graduate qualifications. If it can be done in rural Africa it can certainly be done in Europe.
- Mobile learning academic summaries for PDAs. This has already been implemented by NKI in Norway.
- Mobile learning academic summaries for smartphones. This has been implemented.
- Mobile learning academic summaries for mobile phones. This has already been achieved particularly by the Faculty of Education at the University of Pretoria in South Africa. Extensive numbers of mobile learning for academic summaries have been sent out to teachers in country schools doing post-graduate qualifications. If it can be done in rural Africa it can certainly be done in Europe.
- Full modules by mobile learning for PDAs. All the problems with this have been successfully solved by NKI in the *From e-learning to m-learning project.*
- Full modules by mobile learning for smartphones. This has been successfully achieved in the *Mobile learning: the next generation of learning* project.
- Full modules by mobile learning for mobile phones. The achievement of this will await the full availability of 3G technologies.

CRITERIA FOR INCLUSION IN THE MAINSTREAM

There are four criteria for the inclusion of mobile learning in mainstream education and training. These are:

Enrolment of mobile learning students in courses on the institution's official prospectus. This is essential for incorporating mobile learning into the mainstream. If the mobile learning course is not included in the institution's prospectus and listed as available for student enrolment it remains peripheral with the status of a research project and cannot be considered as part of mainstream provision.

Enrolment of mobile learning students into fee-paying courses. This is essential for incorporating mobile learning into the mainstream. This is applicable to countries in which fees are payable for enrolment in further and higher education courses.. If the mobile learning course is not included in the institution's fee-paying courses and is listed as available for paid student enrolment it remains peripheral with the status of a research project and cannot be considered as part of mainstream provision.

Enrolment of mobile learning students into assessed courses. This is essential for incorporating mobile learning into the mainstream.. If the mobile learning course is not assessed with the same rigour and procedures as other courses offered by the institution it remains peripheral with the status of a research project and cannot be considered as part of mainstream provision.

Enrolment of mobile learning students into accredited courses. This is essential for incorporating mobile learning into the mainstream. As happened in the field of distance education and then in e-learning, the achievement of accreditation for mobile learning is an indication that the sector has entered into the mainstream. The awarding of assessment at university level in distance education was a lengthy process lasting nearly 100 years and was not fully accepted until the foundation of the Open Universities in the 1970s; in elearning it was achieved much more quickly.

PLANNING FOR MOBILE LEARNING

The choice of devices and technologies for mobile learning will depend on the definition of mobile learning adopted.

In this study mobile learning is defined as 'the provision of education and training on mobile devices: Personal Digital Assistants (PDAs), palmtops and handhelds and on smartphones and mobile phones'.

This definition reflects the tension in the field of mobile learning between functionality and mobility. The devices available may be assembled on a continuum running from the most functional to the most mobile. Because the focus in mobile learning is on mobility this presentation limits the range of mobile learning limits the range of the field to the devices listed, to the exclusion of laptop computers.

There are five dimensions to this chapter:

- Definition of devices to be used
- Definition of technologies to be used
- Development of tools and environments to be used
- The search for a wireless Learning Management System (mLMS), that is the changes that are needed to an LMS to 'mobilise' it
- Definition of courses to be developed.

Definition of devices to be used

When we look at how our students and teachers are working today, we find good readability to be of great importance. The materials that we offer online as well as answers to assignments and the assignments themselves, are quite extensive and demand a readable screen. The size is also an issue, but we must also take ease of mobility into consideration. If the screen is to be large, the device will inevitability have to be bigger.

The hypothetically perfect device would be small and fit easily into one's pocket. The screen should fold out to A4 paper size and have paper readability. Wireless connectivity should be of high speed, the user should be always online with the possibility to switch seamlessly between wireless zones and phone networks. The device should have an integrated phone and support all the major office formats for reading and writing as well as pdf format. Security should be high, and if the device is lost the data should be made useless with no risk to the owner. The perfect device should render standard web pages perfectly and offer the ability to strip out advertisements etc. and display useful content only. The web pages should be readable offline as well as online. This should facilitate an understandable on-the-fly text-to-voice and voice-to-text feature.

It is believed the above would be a close to perfect device, but since a device satisfying these specifications is not presently on the market and will not be within the foreseeable future one needs to restrict what one wants and attend to what one needs.

The following list describes a close-to-perfect handheld device:

- Always online connectivity
- Bluetooth for connection with other devices
- Built-in video cannon for displaying presentations etc.
- Camera for documentation in the field
- Flash support
- Full size keyboard available
- Full WI-FI connectivity
- Large storage capacity (Large is a relative term changing with time)
- Screen of acceptable size and readability
- Large battery capacity
- Messaging client for peer-to-peer communication
- Non-volatile memory for backup
- Phone ability
- Read Adobe Acrobat documents
- Read/write common office formats
- Scanner and printer built in
- Small compact device
- Support multimedia content as well as flash, java and java script etc.
- Synchronize and check e-mail with common mail clients.
- Text-to-voice screen reader and Dictaphone

What do we need and why?

The main focus would be on the students' and teachers' needs. These are summed up as:

- A readable screen,

- The ability to read courses online as well as offline and
- Be able to communicate with other students and
- Teachers from wherever they are.

A Personal Digital Assistant (PDA) with communication possibilities can serve these needs. Students and teachers will have easier access to online services and give the teacher the possibility to answer questions while away from his/her personal computer. We wish to allow our students to be able to study on their own when and where they want to.

Recommended requirements for the device:

- Always online connectivity to facilitate synchronous communication and possibly generate quicker response from teachers on e-mail
- Bluetooth for connection with other devices
- Near to full size keyboard to write papers and answer assignments
- Long lasting battery
- *Memory:* 64MB RAM + some (32MB?) ROM for non-volatile memory to have enough space for multimedia content and backup of important documents
- Messaging client for peer-to-peer communication (GPRS would enhance this functionality by being always-online compared to logging on by WLAN or dial-up)
- Read Adobe Acrobat documents
- Read/write Microsoft office Word, Excel, PowerPoint
- Readable screen of good quality
- Small and compact device that fits into one's pocket
- Support multimedia content as well as Flash, Java and java script etc.
- Synchronize and check email with Microsoft mail clients.
- WI-FI connectivity

The Devices Chosen for Proof-of-Concept (PDAs)

Based on the list of what is considered to be needed for a good test of the always-online environment, one can choose to tryout the HP Ipaq with WLAN capability. For testing we will apply both new PDAs with integrated WLAN. The older versions of PDAs are equipped with a jacket that enables the use of a WLAN CF-Card which simulates the always-online environment in the homes of the students who will test the courses and features of the PDA.

Definition of devices to be used (smartphones and mobile phones)

The intention of mobile learning is to utilise the capabilities of existing GSM, GPRS and UMTS mobile telephone handsets and PDAs to increase access to training courses.

It is proposed to develop and trial mobile learning courses for current mobile telephone handsets. Newer 3G (UMTS) devices on the market at phase

2 of the project will be trialled when readily available. Specifically, the devices chosen are the SonyEricsson P900 and the SonyEricsson T610.

P900 Specifications

The Sony Ericsson P900 is a triple band mobile phone, an MP3 audio and MPEG4 video player, an advanced handheld device and a mobile video and digital camera. The phone has e-mail access to a network of contacts. The calendar can be updated. The P900 supports a range of messaging types which will be used to monitor progress and test students. Pictures can be taken with the P900 built-in camera which allows pictures to be received and sent. Text input is facilitated by an on-screen keyboard and a large touchscreen.

Some of the more relevant P900 features are listed below.

- 208x320 pixel display
- Colour LCD 65,536
- MMS (Multimedia Messaging)
- MMS Video
- Video player
- Video streaming
- MP3 Audio
- MPEG4 Video
- Email
- Java
- SMS long (Text Messaging)
- WAP 2.0
- WTLS
- CHTML

Sony-Ericsson T610

The T610 is a sophisticated camera phone. It has a quality 65536 color display. It does not, however have as large a screen size as the P900. Extensive picture sharing options and the latest messaging functionality is available. The T610 supports Java™ download of mobile applications. Some of the more relevant T610 features are listed below.

- Color Display (65536 colors)
- 128x160 pixels display
- e-mail
- GPRS
- Java applications
- Multimedia messaging (MMS)
- WAP 2.0

Further uses

The target groups are IT users and regular students. The first group is able to purchase the latest devices whether PDA-s or smart-phones which regular students may not always afford. For this latter group mobile phones are more familiar devices than handhelds. The assumption is that the common devices suitable for both groups are smartphones. These are not widespread today, however, the dynamic development of the market allows us to forecast their spread in the near future. Considering the advance of technology, we chose one of the multi-purpose devices with the most advanced features available today, the Sony Ericsson P900. The functionality of this device is near to the PDAs, the screen size is almost the same and it has Internet-connectivity in contrast to a regular PDA.

Didactic and evaluation use

Thus development of mobile learning will focus on the Sony Ericsson P900 (smartphone) and the Sony Ericsson T610 (mobile phone).

These devices will be used for two purposes:

- Didactic analysis
- Project evaluation.

For didactic analysis the devices will be used to access and analyse all course materials developed in the project. The analysis will focus on the student userfriendliness of the learning materials developed and feedback will be provided to the course developers on:

- Pedagogical issues
- Screen layouts
- Sequencing of content
- Questioning techniques: SAQs, TMAs, CMAs
- Feedback to the student
- Student to student communication
- Student to tutor communication.

Further analysis will focus on the status of the course materials developed. This is regarded as vital if mobile learning is ever to take off as a viable sector of training provision. The devices will be used to evaluate the products of the project from four perspectives:

- Student userfriendliness
- Didactic efficiency
- Technical feasibility
- Cost effectiveness.

When assessing these criteria for success the evaluation will take into account the needs of mobile learning as a sector of education and training provision.

DEFINITION OF TECHNOLOGIES TO BE USED

How to design for handheld devices?

One needs to design and develop technology that enables a device independent system. The chosen technology should be well adapted and commonly accepted. Hyper Text Mark-up Language (html) is such a standard To specify the look and feel of the pages we already use XSL Transformations (xslt) on the server to output the html based on data from Extensible Markup Language (xml) and look and feel from Extensible Stylesheet Language (xsl). We also use Cascading Style Sheets (css) for layout and design. We are also looking into the newer version of Cascading Style Sheets

There is a media-type designed for print that could give us an improved printer-friendly version of our pages and better layout for small screens. When we implement changes on the server-side we know it will benefit all our clients, the only drawback at the moment is the minimal support for CSS2 by current browsers but support for CSS2 will certainly increase. CSS2 is a World Wide Web Consortium (W3C) standard and they are already working on developing the next generation of CSS. There are several issues defining the usability of a web page, one of these being download time. This is a crucial issue and should be a concern for every type of web page, whether it is meant for reading online using a traditional screen or a PDA (or smart phone). The limited bandwidth on a mobile network is clearly an issue, but the technology is rapidly advancing. Good and efficient coding will result in better utilization of bandwidth and processing power and the use of CSS will reduce the amount of mark-up needed for defining look and feel.

Another issue is the use of pictures, illustrations and graphics in courses. Most of the pictures and illustrations used on web pages for educational purposes are there for a defined reason, not only for navigation and showing pretty pages. This gives us a challenge. How do we display an illustration of a certain size that is needed for the student to understand the learning materials and reach learning objectives? If we keep the size of the picture, the download time will be an issue, but so will screen size. The screen is most of the time limited to 3.8 inches for PDAs and even less for smart phones. We feel that to read and work through an extensive amount of learning materials a screen smaller than the PDA is unacceptable. For smaller courses and on-demand learning or information collection, it could perhaps be acceptable, but not preferable. Some people also turn off images in their browsers and they should get a good description in text format, at least an idea of what they are missing, alt-text is a tag used for this.

There are some PDAs that have larger screens and some web browsers can use the screen in landscape modus which gives a better feel of the page (ThunderHawk) which also uses all of the screen size for content which further enhances the browsing experience.

Evaluation of use of technologies

All the technologies, standards and specifications used in the project will be evaluated on the four criteria listed above:

- Student userfriendliness

- Didactic efficiency
- Technical feasibility
- Cost effectiveness.

The focus throughout will be pedagogical - the contribution of the technologies, standards and specifications used to the creation of a learning environment which will contribute to successful learning for the student. The rules for the creation of successful learning environments for distance education are well known by now, and will be applied to the products of the project.

Cost effectiveness is another major factor. Mobile learning will never take off if the systems designed are not cost-effective for the student/user and in development time, access time, costs for data and voice transmission - all are required to be controlled by the project.

Technical feasibility, especially from the point of view of the student/ user is another important criterion for the success of mobile learning.

Student userfriendliness is vital - from the first mlearning projects there is plenty of evidence, both from students and partners questioning the userfriendliness of studying from smartphone and mobile phone screens.

PRINCIPLES OF THE OPEN MOBILE ALLIANCE

This section looks at the principles of the Open Mobile alliance as they affect mlearning. The principles encourage competition through innovation and differentiation, while ensuring the interoperability of new and existing mobile services across the entire value chain. As you can imagine, this philosophy has major ramifications for the work of mlearning as adherence to OMA approved standards will result in access to the products via any mobile device or system that in turn adheres to the standards of the OMA.

The principles of the OMA are stated as follows:

- Products and services are based on open, global standards, protocols and interfaces and are not locked to proprietary technologies
- The applications layer is bearer agnostic (examples: GSM, GPRS, EDGE, CDMA, UMTS)
- The architecture framework and service enablers are independent of Operating Systems (OS)
- Applications and platforms are interoperable, providing seamless geographic and inter-generational roaming.

The OMA is organized into a number of Working Groups and there are a number of these groups that mlearning will reference quite closely. The first of these is the Browsing and Content Working Group as this group will define the key specifications that mlearning will use in content production. The work of a second group, the Messaging Working Group, is also of great importance to the work as this group will define key issues to do with Multimedia Messaging and Instant Messaging which are two key technologies that will be utilized in mlearning scenarios. Browsing and Content Working Group BAC is specifically chartered to be responsible for base content types, including the semantics and such user agents, behaviour and programming interfaces as is necessary to use such content types, render them and interact with the browser user agent, with the intention of enabling the creation and use of data services on mobile hand held devices, including mobile telephones, pagers and PDAs.

Messaging Working Group

The OMA Messaging Working Group is responsible for the specification of messaging and related enabling technologies. The goal of Messaging Working Group is to specify a set of basic messaging features that may be used to enable specific messaging paradigms.

The Messaging Working Group is also expected to provide clarity of methods by which the messaging enablers are used as a medium for the interaction with different mobile applications.

The scope of the Messaging Working Group is the specification of messaging protocols and features for the Open Mobile Alliance as a whole, dependent on work that may be carried out by other workgroups and other organizations.

The work of both of these working groups will be monitored to ensure industry standard specifications and technologies are utilized. This will ensure that the end products are available to as large a catchment group as possible.

STUDENT EVALUATION OF MOBILE LEARNING

In any new or experimental form of education or training provision, students' evaluation of the new area in real study situaltions has great importance.

A joint questionnaire was developed and administered to the 49 students involved in the evaluation in Ireland, Norway and Hungary. Questions 1-5 asked uestions about the students' personal background: occupation, age, gender, level of education and ownership of mobile devices.

The four dimensions which are crucial for the success of any new training methodology:

- Student user friendliness
- Didactic efficiency
- Technical feasibility
- Cost effectiveness.

The problems with and the objections to mobile learning are well known. The major ones are usually ow memory, insufficient bandwidth, difficult device input and small screen size. It is important to realise that the advance of wireless technologies and the introduction of 3G wireless technologies will quickly improve memory and bandwidth and will ultimately eliminate these as problems. Small screen size is different. It will be a permanent feature of mobile learning and so, to a certain extent will the difficulty of device input though this may be solved by the development of voice input. The reason why small screen size will be a permanent feature of mobile learning comes rom the nature of mobile learning: the provision of education and training on devices which can be carried in one's pocket or in a handbag.

The Norwegian group comments (Dye and Rekkedal 2005) that the test group consisted of 18 employees of NKI. The members of the test group had all volunteered to participate in the trial. They were all registered as students in the test course three weeks before the testing started. Consequently, they had the opportunity to get acquainted with the course on their PC before testing, reading the materials, trying the interactive materials, listening to sound materials and watching video materials before the test on m-learning equipment and solutions started. Largely this is much in line with the assumptions underlying our m-learning developments; i.e. that the students normally study the course from printed materials and their desktop/laptop PC and apply PDA/mobile phone when on the move.

Another advantage with this composition of the test group was that they all had experience with distance education from different positions, insight in distance education pedagogy, problems and challenges.

No member of the test group had in any way been involved in any developments related to the m-learning project. The 3 managers were the personnel director of the NKI group, the director of NKI distance Education, and the director of the Learning Materials Development Department. The non-technical employees were secretaries (2), student advisors (2), editors (2) a salesman and one representative from the marketing department. The technical employees were 2 consultants from the central IT department, one IT consultant and one IT technical assistant from NKI Distance Education.

The group is older than the average distance education student, over 50 per cent being over 50 years. The group also have a higher level of education than the average NKI distance education student.

Concerning mobile device ownership, all test persons, naturally, owned a mobile phone, while only 5 also owned a PDA. Of the 5 test persons who had their own PDA, two had actually stopped using it (because of battery problems) some time ago, and one had newly got a PDA and consequently was not an experienced user.

Concerning our assumptions for the m-learning developments, which is that m-learning functionalities is an additional offer to online distance students who are users of mobile devices and thus may apply this type of equipment when on the move, the test students were at a disadvantage, as most of them needed an introduction to the use of the devices as part of the test. However, this did not turn out to be a problem. Most participants found the equipment quite easy to use.

The test was carried out under a sort of laboratory situation. The test was administered by two of the researchers on the project. One functioned also as tutor during the tests including correction and feedback on test assignments for submission and discussion in the course forum with students. The assistant researcher functioned as observer and guided the test group in using the devices when necessary during the test. Both students and teachers were placed in the same room. This made it possible to observe and register problems and opinions during the trials.

The test was carried out in an "always-online environment" with HP iPAQ h5550 PDAs with built in wireless LAN connected with the NKI Internet College Server. One test person used her newly acquired Qtec 9090 Smart Phone. The iPAQs functions well with Bluetooth and mobile phone connection, while the Qtec has a built-in mobile phone. Phone connection with the server was not included in the tests. This functionality was tested in our previous m-learning project.

The Irish group commented (Flanagan 2005) that during this trial the WCDMA RAN Overview course was adapted in accordance with recommendations for mobile device development. The trial ran from February 2005 to March 2005 during which nineteen people from Ericsson completed the course on Sony Ericsson P910i smart phones and filled out post-evaluation forms.

The evaluation method takes feedback from the overview course under five separate headings. An understanding of the participant's personal background. The participants' evaluation of the user-friendliness of m-learning. The particular interest to this study as the technology and methods used to design this m-learning course have resulted with considerable increases in the didactic efficiency perceived by the participants. Further insight into the participants' evaluation of the technical feasibility of this module; the results from this section again show noticeable increases from previous trials.

The CUB staff also prepared an off-line version of the course material. As the cost of the mobile-internet is quite high in Hungary, the teaching materials were put on the user's mobile devices before the course has started. The html based material could be accessed locally, but this version was lacking of some the functions which were useful in the on-line version.

As the course was about art and as it was developed for mobile devices, the place of the test was in the Hungarian National Gallery, which is located in the Royal Castle in Budapest. The date of the course was the 20th of February, 2005. The CUB staff allocated 2 hours for the whole event, which also included short introduction about the course material and about the test circumstances. The preparation time of the equipment for the event wasn't included. Altogether 12 people tested the course material. 10 of them were students and 2 of them were lecturers. The CUB staff provided the technical assistance for the students, as some of them didn't have any previous experience in using Smartphones or PDAs.

The CUB staff provided 3 Smartphones (Ericsson P900s), 3 PDAs (HP Ipaqs), and a small portable computer. The PDAs didn't have internet connection, so only the off-line course material was accessible, which was stored in the memory of the devices. The P900s were capable of accessing the on-line version as well, but the off line version was available on their memory sticks as well.

The course was fully recorded. The questionnaire consists of two main parts. The first part tries to find out the user's opinion about the course in general, the 2nd part asks questions about the Corvinus specific features.

Altogether 12 people participated the test. (2 trainers, 10 students) That's why the average age of the participants is quite young. All of the people are well educated, 7 of them has already 1-3 years post secondary education, and the other 5 completed 4-5 years in an college or university. All of the participants possess a mobile-phone.

DEVELOPMENT OF COURSEWARE FOR MOBILE LEARNING

This chapter gives the specifications of typical Personal Digital Assistants (PDAs), smartphones and mobile phones of today and discusses the development of mobile learning courseware for them.

PDAS

The Compaq Ipaq 5000 series may be taken as the state of the art for Personal digital Asistants (PDAs) or palmtops or handhelds at the time of writing. Typical of the series is the 5555 which is illustrated below:

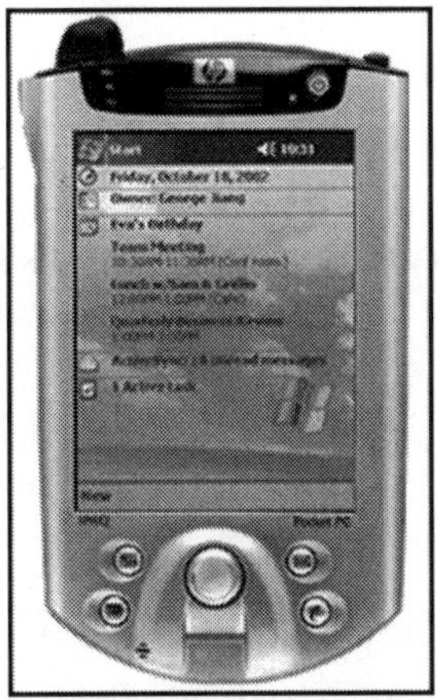

Fig. HP Compaq iPAQ 5555

An update to the iPAQ 5455, the iPAQ 5555 features the new 400MHz PXA255 XScale processor with a 200MHz system bus. This doubles the rate of data flow between the processor and the system memory and increases performance considerably over past XScale units. Other updates include the new Pocket PC 2003 OS - Windows Mobile 2003 for Pocket PC - and 128 MB of RAM. It is claimed that the iPAQ 5555's myriad features make it the most advanced Pocket PC available today.

The HP iPaq 5555 Pocket PC sports a broad range of applications, integrated wireless capabilities, enhanced security and expandability. Continuing to deliver expansion capabilities, the powerful and flexible HP iPAQ Pocket PC 5555 is designed to meet your handheld computing needs with enhanced fingerprint security and wireless connectivity both in and out of the office. One can gain competitive advantage through increased mobile productivity that allows you to communicate, access, and manage information when one wants.

Components

With broad choice of wireless capabilities such as integrated Bluetooth and WLAN 802.11b, one can access the Internet, email, and corporate data at home, at work or on the go.

Versatile expansion capabilities and accessories allow one to extend the HP iPAQ Pocket PC's capabilities or convert it into a unique handheld device. Easily print documents, email attachments and pictures directly from the HP iPAQ Pocket PC to an HP printer. A removable battery, transflective display, integrated Secure Digital slot, increased memory, and Microsoft Pocket PC 2003 Premium Edition give greater productivity with optimal performance.

The HP 5555 allows one to gain competitive advantage through increased mobile productivity that allows one to communicate, access, and manage information when you want.

MOBILE LEARNING COURSEWARE

A mobile device with these specifications is an excellent device for mobile learning. In the *From e-learning to m-learning* project in Norway courseware was developed using Microsoft Reader Works software to produce a comfortable didactic environment for mobile learning students. All students were provided with a PDA with Microsoft Reader software and studied a course for which the content and background readings measured 1000 A4 pages.

In questionnaires filled in by the students who studied this mobile learning course, satisfaction was expressed with the didactic environment created, the students expressed themselves comfortable with the size of the screen and satisfied with reading course text from the display developed with Microsoft Reader. This is of great importance as one of the criticisms frequently levelled at mobile learning is the problem of screen size and forecasts that students would have difficulty studying from the screen of a mobile device. These problems have been shown to be resolvable by the developers of the *From e-learning to m-learning* project courseware in Norway. These Norwegian developers state 'we chose a course previously developed for Internet/web based learning, '*The Tutor in Distance Education*' as courseware for this project. This course is one of our many ongoing Internet courses and therefore already available in a HTML version. Thus it was relatively easy to adapt the existing version of the course to the iPAQ since MS Internet Explorer is the browsing tool used. The main part of the adaptation was to create directories and file structures that insured that all content was present and worked as intended. Some modifications had to be done, e.g. the table of contents had to be changed, so that all links to introductions, study units, articles etc. could be placed on one page.

The Contents page also contains links to examples of course pages such as class list, forum page, the student's personal college page and others. Students can also link to the presentation of their tutor with contact information. The course includes reference links to many external resources, which also are available on the Pocket PC, but accessible only when online through the mobile phone. The course also includes a number of articles available at the Internet College pages. We chose to include the whole library of distance education research reports, articles, conference papers etc. available on the pages. This was done mainly because the course content concerns distance education pedagogy and didactics, thus as the storing capabilities of the iPAQ was sufficient, we considered this as an extra academic service'.

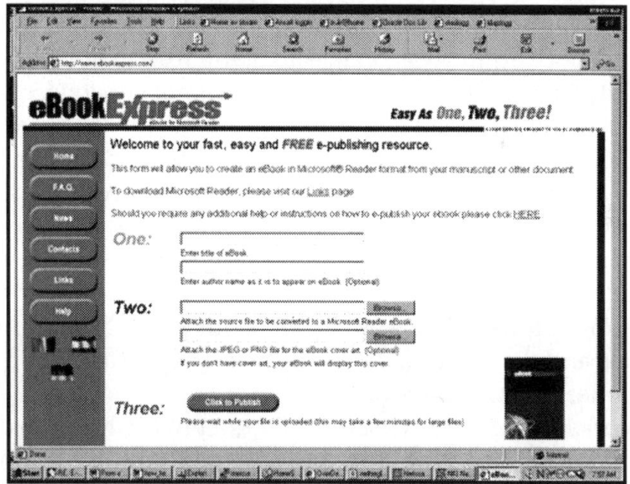

The HTML version applies Microsoft Pocket Internet Explorer that is a web browser withfar less functionality than the full scale PC version.The other version developed for the project uses Microsoft Reader on the Pocket PC as 'browsing' tool.

Microsoft Reader with ClearType is one of the programmes available to read e-books or content in the *.lit file format (MS Reader file format). Microsoft has developed Microsoft Reader with ClearType that enhances display resolution by as much as 300 per cent by improving letter shapes and character spacing, making them appear more detailed, more finely crafted, and more like printed fonts. This gives powerful digital advantages like integrated dictionary support and electronic annotations, while honouring the best traditions of typography to ensure proper kerning and leading, correct margins, and line justification, to name a few. The software also gives the opportunity to read e-books, Pocket Dictionaries etc. to downloaded from the Internet and synchronized to the PocketPC via the PC.

There are several methods to produce materials in the Microsoft Reader format. One may create on-the-fly Reader files via publishing websites like eBookExpress:

The eBook Express Home Page

It is possible to outsource the entire or parts of the converting process. Several e-book consulting and content conversion services are available and offering services ranging from document conversion to complete e-commerce solutions. Overdrive is one example of a firm that offer ePublishing solutions. The software builds the e-book, page-by-page, according to individual preferences to suit the device one is using.

One may also download software that converts publications into Reader files/e-books according to individual preferences. One of these is ReaderWorks. This is a third-party software recommended by Microsoft developed by OverDrive Inc. ReaderWorks is available in three versions, Standard, Publisher and Professional. The Standard version is freeware with less functionality than the Publisher and Professional versions. OverDrive and Microsoft also provides a software development kit (SDK) that software developers can use to build tools that generate Micorosoft Readerfiles. Microsoft also offers an add-in functionality for Microsoft Word that makes it possible to convert a Word document to Reader format.

We have in the project produced a version of the learning materials for Microsoft Reader using the Standard version of ReaderWorks from OverDrive Inc. This version is a freeware application with some limitations regarding commercial sale and distribution. It also lacks the opportunity to provide cover pages and marketing information.

ReaderWorks Standard includes tools to convert html, text and image files to Reader format. It also allow for making a table of content based on heading formatting of HTML documents. Our experience so far is that this is a very well functional tool that also is quite easy to use. It has an intuitive user interface with many different options and functions. We had some problems with empty meta-tags that made the conversion fail. The software also showed some problems with documents containing internal style-sheets and script language. These errors caused the conversion to fail. The HTML code causing these errors had to be manually corrected. The software supplied good reports on what kind of errors arising and where they occurred.

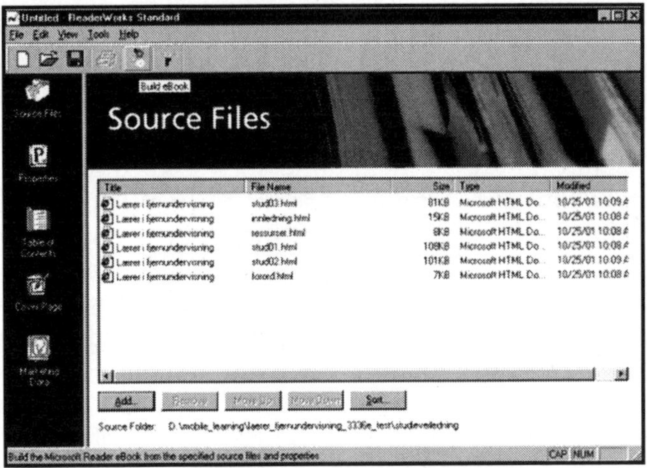

Fig. Screen shot from ReaderWorks software

Student reports on reading and studying on the PDA

The Norwegian researchers report 'All students agreed that reading and studying the learning materials in Microsoft Reader format on the PDA was quite acceptable, specifically related to the ease of carrying the equipment when on the move.

One student wrote that *"In terms of size of the equipment this has been a major advantage compared to for example carrying a laptop around. When travelling to Africa I brought a small makeup bag containing the PDA, Stowaway keyboard and mobile phone. I could fit it in my handbag, which is very convenient. I wished the adapter for phone and PDA was the same to avoid bringing two, but this is of course a minor issue."*

The students point out that the PDA has been very practical as a study tool. They have used it for note taking and also used the text manipulation functions of the Microsoft Reader software.

While the course presented the study materials to be downloaded and studied offline, sending and receiving messages and reading Forum archive entries and contributing to the Forum had to take place online. Very few problems were detected concerning setting up the PDAs for connection to the Internet via mobile phone.

The conclusion is that communicating through the mobile technology and networks functions generally without problems at acceptable speed and costs'.

All the participants are consistently on the positive side. They find the equipment easy to use, they found the experience to be 'fun' and they would be willing to take another m-learning course, and they would also recommend others to take an m-learning course. During the first trial we found that one participant, man between 50 and 60 years, to be quite negative. In this second trial we had one woman at the same age, who was far more positive.

One of the three students is uncertain whether m-learning increases quality of e-learning. The same student is also uncertain whether downloading was easy and whether communication with the tutor was easy. From the whole evaluation process it was clear that this student had some difficulties. From conversations with the student these difficulties seemed to be more related to external factors, time stress and an accident causing sickness absence for a long period during the trial. Generally, the three students in this second trial were somewhat more positive also concerning didactic efficiency.

The students found navigation easy. They do not agree whether graphics and illustrations are necessary. Concerning evaluation and questioning, one did not answer and one is uncertain. We would ascribe this uncertainty to the fact that when the final questionnaire was distributed, they had studied only part of the course and not submitted more than 1 or 2 of the 5 assignments in the course.

One student wrote: *The 'mobility' of the course using the PDA acts as an extension to the general (traditional) online learning programmes. In my case it has been particularly useful when travelling. The slowness of the phone has made me choose to use internet/e-mail with the PDA as little as possible and only when travelling. It may be worth mentioning that I do not have a computer at home, thus the PDA has been useful both for studying materials as well as writing assignments.*

Dye and Fagerberg (2005) from Norway give this advice on developing course materials for PDAs:

In course development one should have focus on having one source for the course content. This means that one should be able to publish the course in different formats from one source. These format could for instance be pdf, (x)html, wml, xml etc.

We urge other course developers to consider whether or not the courses should be developed for the PDA, adapted to the PDA or if the publishing framework should adapt the course to the PDA. We would strongly recommend course materials to be adapted by the framework. Many institutions offer comprehensive courses to their students and the PDA should be considered as a good supplement to ordinary distance education methods. One must take cost/benefit considerations when developing or adapting course materiel to the PDA This goes especially for the use of multimedia elements that often are very expensive to develop and not always suitable for a small screen.

If the course is to be developed for the PDA the main points to remember is that the screen is very small and illustrations should be clean and not full of information. Split an illustration into smaller parts of which a user can get more information as (s)he goes along. It is also important to remember is that reading text on a small screen could be more tiring for the eyes than on a traditional computer screen when reading over a long period of time. One should use shorter paragraphs to improve the readability and for instance use formatting such as bold to emphasize important parts of the text.

SMARTPHONES

Smartphones also have good possibilities for mobile learning course developoment and at 62mm x 42mm have screen sizes somewhat smaller than a PDA.

SonyEricsson P910

The SonyEricsson P910 may be considered a state-of-the-art smartphone, that is a telephone that incorporates many of the features of a PDA, at the time of writing.

A description of the SonyEricsson P910 lists the following features:

The Sony Ericsson P910 is the successor to the P900 and the well-received P800 model released in 2003. It's a smartphone that runs Symbian 7.0 OS and UIQ 2.1. The P910 has a touch screen, handwriting recognition, a large color display and an expansion slot that accepts Memory Stick Duo media. It can synchronise to a PC and supports GPRS for Internet access. The P910 has built-in Bluetooth that allows one to synchronise to a Bluetooth enabled PC, use Bluetooth headsets and use the phone as a wireless modem for a computer or PDA.

It boasts a number of improvements over its predecessor: a better display, more phone-like keypad, predictive text Input, larger capacity battery, video capture and a more business-like design. The P910 is a triband GSM world phone that works on 900, 1800 and 1900 MHz frequencies.

The P910 has a straight-lined industrial design and a monochrome finish. The body of the unit is gun-metal colored with silver around the edges and on the flip cover as well as the battery door, which gives it a slick and business-like style. The 208 x 320 pixel color display lives under a removable flip cover..

On the back of the P910 unit, you will find a camera lens and directly below it a very tiny mirror that you can use to frame self portrait photos. The battery door comprises most of the rear of the phone. As with most other

GSM mobiles, the SIM card is located under the battery. There is an external antenna connector next to the lens which is protected by a tiny rubber cover that you can easily remove using the stylus.

On the left side of the unit, one will find a stereo headset connector, a jog dial, IR port and the power button. On the right side of the P910, one will see a camera button that will activate the CommuniCorder application so you can take pictures or videos. Below the camera button is an Internet button that activates the browser. There is also a Memory Stick Duo slot on the right side. The stylus solo is on the top and right of the P910, and the sync and charging ports are on the bottom of the unit.

The basic phone features

There are two modes of operation: keypad flip closed and flip open (or removed). In flip closed mode, you control the P910 like a traditional cell phone using the keypad and the jog dial. The display is not touch sensitive in this mode, which means you can not use the stylus to control the unit. When you move the jog dial, you will see icons scroll by for messages, contacts, call list, calendar and All Apps. You can navigate and activate these applications with your jog dial. When the flip is opened or removed, you can control the phone using the stylus and jog dial for a more PDA-like experience. You can dial calls using an on-screen virtual keypad.

The P910 has a full set of phone features that include profiles, voice dialing, voice answer, call forwarding, call waiting, speaker phone, flight mode as well as personalized ringtones and a picture phone book. The ringtone support includes MP3, AMR, AU, iMelody, MIDI,RMF (Beatnik) and Wave files.

It also supports speed dialing using the keypad and dialing from the phone's address. Conference calls are supported for users who subscribe to that service.

Telephone reception

Call quality is always good, with no static, distortion or dropped calls, even when the signal meter is at 1 bar. Similarly, one is always able to get a GPRS connection. The battery life is excellent when running PIM, Internet, messaging and multimedia applications. There is a claimed 16-hour talk time and 480-hour standby time, and the P910 outlasts most smartphones. It has an array of power-hungry features (Bluetooth, camera, large color display and fast processor).

The PDA features

The P910 has an extensive set of features built into the device that makes it an excellent choice for productivity, connectivity and multimedia. For input in PDA mode, you've got JOT (handwriting recognition) and an on-screen keyboard. The personal information management (PIM) features include the usual: Contacts, Calendar, Tasks and Jotter (for jotting down notes). The document viewer works with Microsoft Office applications and PDF files. The Internet, email and messaging applications take care of your connectivity needs while movie player, image editor and music player provide a playground for your multimedia content.

Wireless Features

Internet- The P910 comes with an integrated browser labeled Internet that does quite a good job of rendering web pages. This built-in browser supports HTML, WML, WBXML, xHTML basic and mobile profile, cHTML, WAP (version 2.0), WCSS stylesheets, SSL/TLS and WTLS Class 1, 2, 3 for secure web pages.

Cache: 15-120K user-configurable with 30K as default. It also has Java support. You can bookmark web pages, save web pages to read while offline, view push content such as stock quotes, weather reports, etc., and you can view web pages in full screen mode. You can use the jog dial to scroll the pages. If you wish to get a browser that supports even more features, check out Opera, which is available for free on Sony Ericsson's web site.

Messaging- The P910 messaging applications support three types of messages: SMS (Short Message Service), MMS (Multimedia Message Service) and email messages. The Message application puts all the messages in one place and provides a centralized account set up and management window. Before you can receive and send messages, you must set up appropriate accounts for each type that you use.

MMS supports GIF, JPG, BMP, WBMP and PNG formats for images, AMR, AU, iMelody, MIDI and Wav for audio messages and MP4 for video messages. Email support includes POP3, IMAP4, SMTP, and SMTP authentication, SSL-encrypted IMAP4 and SMTP for secured emails. You can also receive and send attachments.

Bluetooth- The P910 has a built-in Bluetooth radio that allows you to use the device with other Bluetooth enabled devices and accessories. To turn on/off and pair the device, launch the Control Panel and click on the Connections tab. The Bluetooth connections manager on the device also allows you to set visible/invisible to other Bluetooth devices for security. The P910 easily discovers our Bluetooth PDAs, access point, desktop PCs and Bluetooth headsets. We tested the device with NextLink Bluespoon headset, Logitech Mobile Bluetooth Headset and Anycom HS 700 Bluetooth Headset and found all worked well with the P900. You can synchronize via Bluetooth to your computer as well. We use a Belkin Bluetooth adapter on our PC, and it works well with the P900 Bluetooth. We paired the P900 with a Sony Clie UX50 and used the phone as a wireless modem for the Clie without a problem (it worked with both the Generic GSM and SE T610 scripts and AT&T service).

Multimedia Tools

CommuniCorder- Sony Ericsson's name for the P910 built-in camera that takes both still photos and videos. The VGA.3 megapixel camera takes decent pictures in well lit environments but doesn't deal with the medium to low indoor lighting very well (images have a strong orange cast). The P910 however does come with a number of pre-installed settings for indoor, outdoor, night time, etc. to improve picture quality. The video quality seems to be better than the photo quality indoors and the sound is quite good. After you've taken the pictures and videos, you can view them, transfer them to the Memory Stick Duo or edit them in the built-in Pictures application.

Pictures- This is a built-in image editor that provides functions for managing your photos (rename, view thumbnail, send as, delete, move, copy, etc.) and editing the images. The application is heavy on managing photos, image files and light on editing tools which include only a few drawing tools, cropping ability, resizing pictures and zooming. The image formats supported by this application include JPEG, BMP, GIF (including animated), PNG and WBMP.

Video player- The built-in Video player plays MPEG-4 videos either in the app's window or in full screen mode. You can manage video clips shot with the P900 or imported to the device. Using this app, you can also watch steaming video on the Internet.

Music player- As with the image viewer and video player, if you have music files on either internal memory or the Memory Stick Duo, you will see them in the music player. The built-in music player supports MP3, Wav, AU, AMR and G-MEDI level 1 with 24 voices polyphony, RMF and iMelody files. The MP3 through stereo headset has excellent quality. You can use the jog dial to control the volume. The player interface is a simple one, offering features that allow you to manage music files, see the track list and detailed track info.

There are other applications such as sound recorder, calculator, Clock as well as a very capable File Manager you can explore on your own. It is quite easy and very enjoyable to personalize your P900 by using themes, wallpapers, screen savers and ringtones.

5

Mobile and Cellular Standards Process

INTRODUCTION

The standards organizations and partnership projects provide technical input to the global standards developer for ratification and approval.

The International Telecommunications Union (ITU), a charter organization of the United Nations, is the pre-eminent global standards developer for telecommunications. The ITU-R (radio communications sector) addresses terrestrial and space (satellite) radio communication. Standards development organizations and partnership projects, listed below, provide technical input to the ITU for ratification and approval.

CELLULAR HANDOVER AND HANDOFF

The concept of a cellular phone system is that it has a large number base stations covering a small area (cells), and as a result frequencies are able to be re-used. Cell phone systems also provide mobility. As a result it is a very basic requirement of the system that as the mobile handset moves out of one cell to the next, it must be possible to hand the call over from the base station of the first cell, to that of the next with no discernable disruption to the call. There are two terms for this process: cellular handover is used within Europe, whereas cellular handoff is the term used in North America.

The handover or handoff process is of major importance within any cellular telecommunications network. It is necessary to ensure it can be performed reliably and without disruption to any calls. Failure for it to perform reliably can result in dropped calls, and this is one of the key factors that can lead to customer dissatisfaction, which in turn may lead to them changing to another cellular network provider. Accordingly handover or handoff is one of the key performance indicators monitored so that a robust cellular handover/ handoff regime is maintained on the cellular network.

HANDOVER BASICS

Although the concept of cellular handover or cellular handoff is relatively straightforward, it is not an easy process to implement in reality. The cellular network needs to decide when handover or handoff is necessary, and to which cell. Also when the handover occurs it is necessary to re-route the call to the relevant base station along with changing the communication between the mobile and the base station to a new channel. All of this needs to be undertaken without any noticeable interruption to the call. The process is quite complicated, and in early systems calls were often lost if the process did not work correctly.

Different cellular standards handle hand over/ handoff in slightly different ways. Therefore for the sake of an explanation the example of the way that GSM handles handover is given.

There are a number of parameters that need to be known to determine whether a handover is required. The signal strength of the base station with which communication is being made, along with the signal strengths of the surrounding stations. Additionally the availability of channels also needs to be known. The mobile is obviously best suited to monitor the strength of the base stations, but only the cellular network knows the status of channel availability and the network makes the decision about when the handover is to take place and to which channel of which cell.

Accordingly the mobile continually monitors the signal strengths of the base stations it can hear, including the one it is currently using, and it feeds this information back. When the strength of the signal from the base station that the mobile is using starts to fall to a level where action needs to be taken the cellular network looks at the reported strength of the signals from other cells reported by the mobile.

It then checks for channel availability, and if one is available it informs this new cell to reserve a channel for the incoming mobile. When ready, the current base station passes the information for the new channel to the mobile, which then makes the change. Once there the mobile sends a message on the new channel to inform the network it has arrived. If this message is successfully sent and received then the network shuts down communication with the mobile on the old channel, freeing it up for other users, and all communication takes place on the new channel.

Under some circumstances such as when one base transceiver station is nearing its capacity, the network may decide to hand some mobiles over to another base transceiver station they are receiving that has more capacity, and in this way reduce the load on the base transceiver station that is nearly running to capacity. In this way access can be opened to the maximum number of users. In fact channel usage and capacity are very important factors in the design of a cellular network.

TYPES OF HANDOVER/ HANDOFF

With the advent of CDMA systems where the same channels can be used by several mobiles, and where it is possible to adjacent cells or cell sectors to use the same frequency channel there are a number of different types of handover that can be performed:

- Hard handover (hard handoff)
- Soft handover (soft handoff)
- Softer handover (softer handoff)

Although all of these forms of handover or handoff enable the cellular phone to be connected to a different cell or different cell sector, they are performed in slightly different ways and are available under different conditions.

Hard handover

The definition of a hard handover or handoff is one where an existing connection must be broken before the new one is established. One example of hard handover is when frequencies are changed. As the mobile will normally only be able to transmit on one frequency at a time, the connection must be broken before it can move to the new channel where the connection is re-established. This is often termed and inter-frequency hard handover. While this is the most common form of hard handoff, it is not the only one. It is also possible to have intra-frequency hard handovers where the frequency channel remains the same.

Although there is generally a short break in transmission, this is normally short enough not to be noticed by the user.

Soft handover

The new 3G technologies use CDMA where it is possible to have neighbouring cells on the same frequency and this opens the possibility of having a form of handover or handoff where it is not necessary to break the connection. This is called soft handover or soft handoff, and it is defined as a handover where a new connection is established before the old one is released. In UMTS most of the handovers that are performed are intra-frequency soft handovers.

Softer handover

The third type of hand over is termed a softer handover, or handoff. In this instance a new signal is either added to or deleted from the active set of signals. It may also occur when a signal is replaced by a stronger signal from a different sector under the same base station. This type of handover or handoff is available within UMTS as well as CDMA2000.

Cellular handover or cellular handoff are performed by all cellular telecommunications networks, and they are a core element of the whole concept of cellular telecommunications. There are a number of requirements for the process. The first is that it occurs reliably and if it does not, users soon become dissatisfied and choose another network provider in a process known as "churn". However it needs to be accomplished in the most efficient manner. Although softer handoff is the most reliable, it also uses more network capacity. The reason for this is that it is communicating with more than one sector or base station at any given instance. Soft handover is also less efficient than hard handover, but again more reliable as the connection is never lost.

STANDARDS DEVELOPMENT ORGANIZATIONS

Standards development organizations (SDOs) are national or multinational organizations, actively involved in defining the next-generation wireless standards, along with refining the ongoing remedial editing of existing standards SDOs are comprised of various companies who work together to promote specification proposals. SDOs also study radio spectrum utilization, including such subsets as intersystem issues, emergency services, and accommodations for the disabled. They also coordinate and cooperate with the ITU on standardization of radio systems in the field of telecommunications. The coordination and cooperation issues are managed by "Harmonization" groups. The following is a list of Western standard development organizations, along with their respective areas of geographical and technical interests:

- The European Telecommunications Standards Institute (ETSI) is defining a technology standard for 3G called the Universal Mobile Telecommunications Systems (UMTS).
- The Japan Association of Radio Industries and Business (ARIB) primarily focuses on WCDMA for IMT-2000.
- The primary Canadian SDO is the Telecommunications Standards Advisory Council of Canada (TSACC).
- The American National Standards Institute (ANSI) is a US repository for standards considered to be semi-permanent, a nebulous term for "longer than interim."
- The United States Telecommunications Industry Association (TIA) and T1 have presented several technology proposals on WCDMA, TDMA UWC-136 (based upon D-AMPS IS-136), and cdma2000 (based upon IS-95). The American National Standards Institute (ANSI) accredits both TIA and T1. The primary standards working groups are TR45 (Mobile & Personal Communications 900 & 1800 Standards and TR46 (Mobile & Personal Communications 1800 only Standards).

The Asian standards development organizations include the Korean Telecommunications Technology Association (TTA) and China Wireless Telecommunications Standards Group (CWTS), Partnership Projects.

The Third-Generation Partnership Project (3GPP) was formed by SDOs and other related standards' bodies to harmonize European, Asian, and North American standards proposals, and to define a complete set of global technical specifications for third-generation mobile systems based upon the evolved GSM core networks and radio access technologies. The project is better known as "3GPP."

3GPP is comprised of the following SDOs: ARIB (Japan), CWTS (China), ETSI (Europe), T1 (USA), and TTA (Korea). The project is divided into several technical specification groups (TSG's), with each TSG having multiple working groups, each responsible for defining an aspect of the third-generation standard. 3GPP is chartered to define the radio access network, the core network (including mobility management and global roaming), terminal access to the network (including specifications for the user identification module), and systems and services. (Additional information is available at the 3GPP web site *http://www.3gpp.org*.

The Third-Generation Partnership Project 2 (3GPP2) was organized by the SDOs that were concentrating on the development and evolution of the American National Standard (ANSI/TIA-41 core networks) and the relevant radio access technologies. The five SDOs are ARIB (Japan), CWTS (China), TIA (USA), TTA (Korea), and TTC (Japan). Similar to 3GPP, 3GPP2 is also comprised of several technical specification groups, each with multiple working groups. (Additional information is available at the 3GPP2 web site *http://www.3gpp2.org*.)

Both 3GPP and 3GPP2 organizations recognize the ITU as the preeminent organization for international telecommunications standardization, and they have agreed to submit all results to the ITU for consideration and approval.

THE MIGRATION TO 3G

The genesis of today's wireless technology began in the early 1980's with the introduction of the first mobile cellular handsets. These systems utilized analog interface technology and supported voice-only capabilities. This technology is still used in many parts of the world; however, it is limited in bandwidth and is low in quality. With the high demand for cell phones and the increased need for enhanced quality and more features, the Second Generation was introduced. 2G is primarily voice only, but does provide higher bandwidth, better voice quality, and limited data services that use packet data technology. 2G systems are currently in wide deployment with enhanced 2G systems currently available on the market. The 3G standardization process is coming to closure, with the recent completion of the IMT-2000 radio interface recommendations.

FIRST-GENERATION MOBILE STANDARDS

The first generation of cellular wireless communications was based on analog technology and progressively became available to the consumer during the late 1970's and early 1980's. The most successful analog systems are based on the following standards, all of which are still in demand today:

Nordic Mobile Telephone (NMT) was the first commercially available analog system, introduced in Sweden and Norway in 1979.

Advanced Mobile Phone Service (AMPS) was launched in 1982. This has proven to be the most successful analog standard of all. AMPS networks are widely deployed and can be found on all continents.

Total Access Communications System (TACS) was originally specified for the United Kingdom and is based on AMPS. The original TACS specification was extended and is known as ETACS. ETACS is primarily deployed in Asia Pacific regions.

Second-Generation Mobile Standards

The second-generation (also known as 2G) introduced digital wireless standards that concentrated on improving voice quality, coverage, and capacity. The 2G standards were defined and designed to support voice and low-rate data only—Internet browsing was in its infancy during the definition stage. The world's four primary mobile digital wireless standards currently deployed around the world are GSM, TDMA (IS-136), CDMA (IS-95-B), and PDC, all supporting data rates up to 9.6kbps.

Global System for Mobile phone communications (GSM) was the first commercially available digital standard, introduced in 1992. GSM relies on circuit-switched data. The basic development of supporting data at low bit-rates (<9.6 kbps) was introduced at the beginning of commercial services and has been predominantly used for e-mailing from laptop computers.

Time Division Multiple Access, originally IS-54 and now IS-136 (TDMA IS-136), is sometimes referred to as the "North American" digital standard; however, it is also deployed in Latin America, Asia Pacific, and Eastern Europe.

Personal Digital Communications (PDC) is the primary digital standard in Japan.

IS-95 is based on "narrowband" (referred to as narrowband because of the limited amount of information that can flow through these networks) Code Division Multiple Access (CDMA) technology. It has become popular in South Korea and North America.

Enhanced Second-Generation Mobile Standards

Enhanced second-generation (sometimes referred to as 2.5G or 2+G) builds upon the second-generation standards by providing increased bit-rates and bringing limited data capability. Data rates range from 57.6kbps to 171.2kbps.

High-Speed Circuit-Switched Data (HSCSD) provides access to four channels simultaneously, theoretically providing four times the bandwidth (57.6) of a standard circuit-switched data transmission of 14.4kbps. D-AMPS IS-136B Time Division Multiple Access (TDMA) is the intermediate step to

Universal Wireless Communication (UWC-136), a third-generation standard. The first phase of D-AMPS will provide up to 64kbps. The second phase will provide up to 115kbps in a mobile environment.

General Packet Radio System (GPRS) is an evolutionary path for GSM and IS-136 TDMA to UWC-136. It is a standard from the European Telecommunications Standards Institute (ETSI) on packet data in GSM systems. The Telecommunications Industry Association (TIA), as the packet-data SDO for TDMA-136 systems, has also accepted GPRS. GPRS supports theoretical data rates up to 171.2kbps by utilizing all eight channels simultaneously. This data rate is roughly three times faster than today's fixed telecommunication networks and about ten times as fast as current circuit-switched data services on GSM networks. GPRS is a universal packet-switched data service in GSM. It involves overlaying a packet-based air interface on the existing circuit-switched GSM network. Packet switching means that GPRS radio resources are used only when users are actually sending or receiving data. Using GPRS, the information is split into separate but related packets before being transmitted and subsequently reassembled at the receiving end. GPRS is a non-voice-added service that allows information to be sent and received across multiple mobile telephone networks. It supplements today's circuit-switched data and short messaging service. GPRS uses packet data technology, a fundamental change from circuit-switched technology, to transfer information. It also facilitates instant connection capability, sometimes referred to as "always connected." Immediacy is one of the key advantages of GPRS. Immediacy enables time-critical application services.

Third-Generation Mobile Standards

Third-generation systems will provide wide-area coverage at 384kbps and local area coverage up to 2Mbps. The primary motivation for the development of third-generation wireless communications is the ability to supplement standardized 2G and 2G+ services with wideband services. Essentially, this offers voice plus data capability.

The existing array of incompatible second-generation technologies, together with the restricted amount of information that can be transferred over these narrowband systems, prompted the ITU to work towards defining a new global standard for the next-generation broadband mobile telecommunication systems. Known as IMT-2000 (International Mobile Telecommunications-2000), the project was started to attain authorship of a set of globally harmonized standards for broadband mobile communications. The first set of IMT-2000 recommendations was recently approved by the ITU.

IMT-2000 is the term used by the International Telecommunications Union for this set of globally harmonized standards. The initiative was to define the goal of accessing the global telecommunication infrastructure through both satellite and terrestrial mobile systems. IMT-2000 has reflected the explosion of mobile usage and the need for future high-speed data communications, with wideband mobile submissions. IMT-2000 is a flexible standard that allows operators around the world the freedom of radio access methods and of core networks so that they can openly implement and evolve their systems. How they do it depends on regulations and market requirements.

The recent IMT-2000 recommendation highlights five distinct mobile/terrestrial radio interface standards:

- *IMT-MC:* CDMA Multi-Carrier (known as cdma2000 or IS-2000).
- *IMT-DS:* CDMA Direct Spread (known as Wideband CMDA or WCDMA-FDD). This standard is intended for applications in public macro-cell and micro-cell environments. The Frequency Division Duplex (FDD) mode is used for symmetrical applications, i.e., those requiring the same amount of radio resources in the uplink as in the downlink. This standard is well supported by Japan's ARIB and GSM network operators and vendors.
- *IMT-TC:* CDMA TDD (WCDMA-TDD). Time Division Duplex (TDD) targets public micro-cell and pico-cell environments, and, due to severe interference-related considerations, is intended primarily for indoor use. This standard is optimized for symmetrical and asymmetrical applications with high data rates.
- *IMT-SC:* TDMA Single Carrier (known as UWC-136 and EDGE). UWC-136 (Universal Wireless Communications) and EDGE (Enhanced Data Rates for GSM Evolution) will provide extended data services, with no changes to channel structure, frequency, or bandwidth. IMT-SC is the evolutionary path for GSM and TDMA-136, achieved by building upon enhanced versions of GSM and TDMA-136 technology. EDGE is a radio-based high-speed mobile data standard with aggregate transmission speeds of up to 384kbps when all eight timeslots are used.
- *IMT-FT:* TDMA Multi-Carrier (well known as DECT, Digital Enhanced Cordless Telecommunication).

The IMT-2000 recommendations encompass three CDMA and two TDMA radio air interface standards.

Wideband Code Division Multiple Access (WCDMA) should not be confused with narrowband CDMA; they are completely different protocols. WCDMA is a younger technology, defined specifically to deliver high-speed data services and Internet-based packet-data at 3G data rates. WCDMA supports both packet and circuit-switched communications, such as Internet access and landline telephone services; however, WCDMA was defined with no requirements on second-generation backward compatibility.

WCDMA makes very efficient use of the available radio spectrum. No frequency planning is needed, since one-cell re-use is applied. Using techniques such as adaptive antenna arrays, hierarchical cell structures, and coherent demodulation, network capacity can be increased. In addition, circuit and packet-switched services can be combined on the same channel, allowing true multimedia services with multiple packet or circuit connections on a single terminal. WCDMA capacity is approximately double that of narrowband CDMA. The wider bandwidth and the use of both coherent demodulation and fast power control in the uplinks and the downlinks allow a lower receiver threshold. WCDMA uses a network protocol structure (signaling) similar to that of GSM; therefore, it will be able to use the existing GSM network as the core network infrastructure.

In CDMA2000, a range of RF channel bandwidths are supported: 1.25, 3.75, 7.5, 11.25, and 15MHz. This range allows for support of a range of data rates as well as a high number of users.

In order to support higher bandwidth channels, CDMA2000 has defined two configuration options: Direct Spread (DS) and Multi-Carrier (MC). The DS option is similar to IS-95B and uses the entire bandwidth to spread the data for radio transmissions. In the MC option, user data is encoded as a single stream and de-multiplexed into multiple streams. Each stream carries part of the user data using a different carrier frequency signal, hence the name Multi-Carrier. The receiver will multiplex the received signals together before demodulation is carried out. Both the DS and MC options are available in the forward link only. The reverse link supports only the DS option.

Time Division Multiple Access

One approach to reducing the number of confusing options to the end user and to improve the overall functionality of time-division cellular technology is to combine TDMA and CDMA radio air interface technology into one system. This combined approach, referred to as TD-CDMA, would retain some of the fundamental GSM-TDMA design parameters, such as frame and time-slot structure, which are key factors for interoperability and evolution. At the same time, the CDMA technology would add better interference averaging and frequency diversity. The combined approach would also merge the excellent spectral efficiency of CDMA, while retaining the robustness, planning principles, and well understood characteristics of TDMA-based GSM. In addition to the improvements of data throughput and interworking, 3G will provide an additional spectrum for the operators. The increase in 3G spectrum efficiency will also provide the operator with more throughput over limited resources. The transition from the existing 2G networks to 3G capabilities will evolve over time. Dual-mode terminals will attempt to provide seamless hand-over and roaming capabilities.

THIRD GENERATION OF MOBILE COMMUNICATION SYSTEM

The implementation of 3G networks in Norway and the other European countries follows the European Telecommunications Standards Institute (ETSI) standard Universal Mobile Telephone system (UMTS). The aim of the UMTS standard is to create one European mobile telephone network enabling users to use their mobile handsets to access high-value services seamlessly in Europe, and finally all over the world. Telecommunication systems have a historically high strategic significance in modern and industrial societies (Godø 1995). This also concerns UMTS in future Europe as part of the information society spelled out by The Commission of European Communities:

"Advanced wireless platforms such as 3G are an essential building block to achieve the goals of the Information Society in terms of consumer demand, productivity, competitiveness and job creation."(EU 2002).

In the scattered populated and topologically complex geography (in terms of mobile infrastructure implementation) of Norway, the governmental telecommunication policy has also historically been seen as fundamental to sustainable regional development, are described as the political aims for UMTS in the hearing of the Norwegian Ministry of Transport and Communications concerning the Norwegian mobile communication market:

"Ensure households and businesses all over the country basic telecommunication services with high quality and low price; Ensure maximum of value creation and efficient utilisation of the resources in the telecom sector by ensuring access and efficient use of the public telecommunication network and services by effective competition."

Telecommunication is in addition seen as a source for industrial and job opportunities, by exporting technology and knowledge to other European and non-European countries (NMTC 1999). The achievement of these stated aims is expected to come by enabling effective competition in a liberalised telecommunication market. In the case of Norway it can be argued that policy makers and the telecommunication operators both have to meet the classical challenges for less developed regions of the world described by McDowell: The expansion of the telecommunication infrastructure due to the geographical challenges, and the challenges for the post-industrial North: the issues of competition, pricing, and social objectives (McDowell 2001), to succeed in implementing UMTS.

The strategic importance of telecommunication is also reflected in the numerous national and international governmental bodies concerned with the use of frequencies, attribution of licences and competition. In Norway, the Norwegian Ministry of Transportation and Communication, the Norwegian Post and Telecommunications Authority, the International Telecommunication Union (ITU) and different EU bodies have the major stake in the current regulation policies. The operators and other stakeholders are also given a voice in numerous hearings. National regulations in telecommunication have in Norway lately undergone changes as to suite the relatively newly deregulated and liberalised European telecommunication market. The deregulation has resulted in an open market, where the monopoly situation of the state owned "Televerket" no longer exists *de jure*. *De facto*, Telenor (formerly "Televerket") still has a significant market position in mobile communication, and together with the second largest operator: NetCom, respectively having 62.5 per cent and 26.3 per cent per cent of subscriptions and 70.2 per cent and 21.4 per cent of the traffic minutes (Statistics for 2001, NPTA 2002a), the market are basically a duopoly. The strong position of Telenor have been an important issue for national regulation bodies, trying to secure the access to the existing European mobile communication system GSM network and the market for new entrants (as for example Sense, the third largest operator having 5.44 per cent of subscriptions and 4.69 per cent of traffic minutes) to create effective competition and correct pricing of the services. Actors (Telenor and NetCom only) with their own physical network have obviously more flexibility than the operators and service providers without. This is acknowledged by the Ministry of Transportation and Communication through obligating licensees to build parallel and completely independent UMTS networks.

Sharing the air

Radio frequencies combined with antennas and transmission effect enables communication over different ranges. The different ranges make some frequencies more popular than other, and they eventually become crowded. Radio frequency regulation authorities are responsible for regulating the use of different frequencies so that communication doesn't break down, as well enabling a maximum of utilization. Regulations are implemented by licensing the use and appropriate the different parts of the frequency spectrum to different purposes, and different licensees. GSM and UMTS are appropriated different bands in the frequency spectrum, fitting the frequency regulation tables of Norway as well as Europe in general. In practice, the licensees them selves are responsible to effectuate sharing policies and avoid disturbance of other licensees and end-users. To ensure this, the very much of the spectrum requires the individual end-users to be certified prior to use. This is not the case of GSM and UMTS, as the responsibility and certification is handed over to the handset manufacturers by certification of products.

Mobile communication system configurations are based on *cells*. The cell is a circular area surrounding the base-stations that is delimited by the radius of reach of base-stations signals. The end-user communicates with other end-users through the base-stations that are connected with fixed cables. To keep users connected as they move between cells, or *roam*, the infrastructure must basically be configured such that cells are overlapping, illustrated by Figure below.

Sharing the air is a challenge when more than one user is communicating in the same cell at the same time. If this happens, and users use the same frequencies, transmissions will brake down. These collisions can be avoided by not reusing frequencies in a cell and in adjacent cells, as frequencies used in cell *c1* can be reused in cell *c8*, but not in the other cells. Mobile communication systems implements in addition more advanced procedures to share the air.

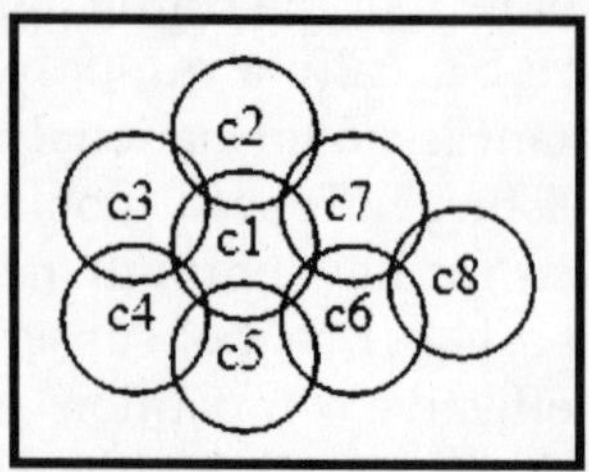

Fig. Principles of cells

GSM uses frequency sharing based on time-slots, implying that users only use a frequency for 0.5 milliseconds before others can use the frequency. Using this schema, the users must be granted new frequency and new time slot each time they switch to a new base-station when roaming, making the *handover* procedure complicated. The smaller cell, the more often handover must be effectuated.

The high density of use in urban areas, and also in more rural due to the high penetration of mobile phones in Norway, requires in addition more sophisticated use of frequency sharing, implemented by UMTS. Instead of sharing the air by division of frequencies and time, UMTS spreads radio signals over a range of frequencies, using CDMA (Code Division Multiple Access). With this schema other concurrent transmission is just seen as noise. This makes the theoretical simultaneous users much higher than with GSM, as well as making the handover procedures easier as the same schema and frequencies are used by any base-stations. At the same time, the bandwidth is highly dependent on the number of users in one and the same cell, as well as which kind of data is communicated. If all users in a UMTS cell are using the lowest transmission capacity (voice), 256 concurrent users can communicate simultaneously. If everyone uses maximum capacity (multimedia), only two concurrent users are possible.

The frequencies used by GSM have a range of maximum 35 kilometers while the UMTS frequencies only reach 6 kilometers, both due to the nature of the frequency band and the bandwidth provided. The range of transmission decides how large *cell* the base-station creates, and the needed density of base-stations. Using different effect in transmission, different cell sizes are implemented in UMTS networks to optimize the available bandwidth on the basis of user density (e.g. macro-, micro- and pico-cells). The bandwidth available will also be affected when the handsets are one the move, and decreases as the speed increases, as high-speed radio access to highly mobile users is difficult. This illustrates the need for a much higher density of UMTS base-stations, at a cost of approximately 1 million NOK each. In addition, the broadband cabled network coupling the UMTS base-stations to the cabled network would probably become as expensive as the base-stations.

UMTS

The European countries have chosen UTMS as the standard for the third generation of mobile communication system. The third generation in the Nordic countries have been preceded by the first generation mobile telephone system NMT (Nordic Mobile Telephone), and the second generation GSM (Global system for Mobile Communication). UMTS is supposed to coexist with 2G for several years, as NMT coexisted with GSM from 1992 to 2002, when NMT finally was phased out. As with NMT and GSM, GSM handsets will not be compatible with the UMTS. On the other hand, the different networks will not be totally separated as UMTS handsets will be compatible with GSM providing a dual mode of communication (This backward compatibility was not the case with GSM and NMT). This is because UMTS, as NMT, will provide less coverage than GSM, probably only rural areas with high density of users.

UMTS is by EU defined as:

"... a third-generation mobile and wireless communications system capable of supporting in particular innovative multimedia services, beyond the capability of second generation systems such as GSM..." (EU 1998, article 2, p4).

This definition is not very specific when it comes to what kind of new services to expect. ITU is slightly more specific as they define UMTS by higher capacity and enhanced network functionalities, which allow advanced services and applications, including multimedia. The major difference between 3G and 2G from the end-user perspective is highlighted by these definitions to be the increased user value created by a broad service offering (EU 2002a). The key features for end-users provided by UMTS are:

- Provision of Internet and other multimedia applications.
- A range of content services.
- Worldwide roaming capability.

UMTS will provide basic services as with GSM, i.e. voice, SMS, roaming and billing. In addition, multimedia services developed by operators as well as service providers will create additional value and it is probable and expected that different forms of service providers will have a more highlighted role in the future, implying new business models and new forms of revenue sharing. The content services that will be provided by UMTS will evolve from the "there and then" voice and SMS services provided by GSM. These services provided by UMTS can basically be described as "mobile Internet", where content is optimised for the limited graphical abilities of the mobile handsets, as well as utilising functionality as location awareness. A range of services have been suggested, as for example:

- Images and video: streaming pictures and video, entertainment, games, lottery, gambling, pornography and video-conferencing.

- Short range and Location based services: Information of nearby restaurants, shops and special offers.
- Context based services, or push: Notification of interesting sports event if you are in front of the TV, advertising if you are close to the advertiser
- Simultaneous transfer of speech, data, text, pictures, audio and video.
- High-speed, mobile access to the Internet.
- Customized infotainment.
- Travel information: congested roads, flight departures. And if you get lost, find your current location.

Roaming

When mobile users are roaming (moving from one area (cell) to another) the handset must switch to the closest base-station. When one base-station hands over the user to another base-station a multiple of transactions must be made. If the user is roaming among base-stations belonging to the subscribed operator, this is only a technical issue. But when roaming implies other operators, national or international, the picture becomes more complicated. The vision of UMTS presupposes that it will be possible for users to roam freely with their handsets and have affordable access to equivalent products and services globally by end-users utilizing the most proper network despite which operator it belongs to. Today, technically and efficient roaming exists for the GSM standard throughout Europe, parts of Africa, Asia and South America. However, it is far from global: for instance, roaming to and from North America and Japan requires special measures and users may even incur custom duties and taxes relating to the use of their handsets abroad. Moreover, the complexity and non-transparency of pricing for roaming services have been problematic for 2G users.

International roaming requires bilateral agreements between the home operator and operators in the foreign country where the roaming end-user is situated. In Europe, this implies more than 200 agreements for each operator, in practise eased using brokers for negotiations. Using foreign operators are especially problematic for the end-users concerning the tariffs. Billing is not done homogenous, and either per minute or per second, "peak" hours where prices are high are different from country to country, as well as the charges are different among the foreign operators, as price variances of ten times are reported (Sutherland 2001).

These charges are also much higher than the local charges, seemingly having no relationship with the underlying costs as well as being relatively to other operators. This has resulted in user organizations withdrawing phones, forbid use abroad and encouraging alternatives (Sutherland 2001). The price structure is also opaque, as some foreign operators don't give the price information to visitors, as well as the home operator adding their own percentage on the tariffs. As home operators makes little effort to inform of the foreign charges as well as the possibility for the user to choose among different operators when abroad, this eventually leads to no competition and prices out of the control. The complexity of pricing for international roaming will become even more intolerable when simple voice services will be accompanied with advanced as well as multimedia services by UMTS. This implies additional pricing structures that possibly will differentiate on time units on simple voice, per use on services as SMS, and amount of data transferred on multimedia services.

UMTS actually steps back when it comes to national roaming, that is well handled by GSM. This is primarily a regulation and effective competition issue, as Norwegian licensees are prohibited to cooperate, as the UMTS networks must appear as independent nationally to ensure effective competition. Thus, national roaming as with GSM, implying automatically usage of non-subscribed network where subscribed operator has no coverage, is prohibited. This is further complicated as UMTS networks at least in nearest feature only will cover urban and densely populated areas, implying that GSM, and not UMTS networks will be available for UMTS subscribers when out of reach of own operator.

Bandwidth

GSM infrastructure provides bandwidth of 9.6 kbit/s for data transfer. Later enhancements using the GSM infrastructure, also called 2.5G, theoretically provide 57.6 kbit/s with HSCSD (High Speed Circuit Switched Data), and 144 kbit/s with GPRS (General Packet Radio Service). In practice, HSCSD today provides up to 28.8 kbit/s, but always at least 9.6 kbit/s, in comparison with GPRS that have large variations, and only 2-4 times the capacity of GPRS. In addition, the bandwidth is dependent upon what capacity the phones offer, currently up to 30 kbit/s). Based on theoretical numbers, without concern of interference and the number of other users connecting to the same base-station, low mobility users, those based in or near buildings travelling at less than 10kmph or stationary can expect 2 Mbp/s from UMTS. Full mobility users, those travelling at less than 120kmph and in urban outdoor environments can expect 384kbps. High mobility users, classed as users travelling over 120kmph in rural areas can expect data rates of 144Kbps. This increased bandwidth is the basic enabling feature for the new services provided by UMTS.

The Standardisation process

The work on the standard for 3G started in the 1985 by the International Telecommunication Union (ITU) within the Interim Working Party 8/13, and work continued in Task Group 8/1. ITU got its name in 1932, when the International Telegraph Convention of 1865 and the International

Radiotelegraph Convention of 1906 was combined, and in1947 it became a UN specialised agency. Lately, the union has focused on strengthening the participation of the private sector in their work. The Norwegian Post and Telecommunication Authority represent Norway in ITU on the behalf of the Norwegian Ministry of Transport and Communications. The ITU initiative resulted in the global IMT-2000 standard later adopted by the European Telecommunications Standards Institute (ETSI) in 1991.

ETSI formed the SMG5 group with the mandate to form UMTS. In other parts of the world, as the US and Asia similar local efforts started. After pressure from ITU to harmonize the different standards that emerged from these efforts, the Third Generation Partnership Project (3GPP) was established in 1998, with its mission to produce globally applicable technical specifications, also having the responsibility to formally specify UMTS.

In the European countries the planned rollout as faced delays, is both explained by the economical downturn after the operators where licences, as well as delays in the standard and the equipment (NMTC 2001). Base-station equipment has been developed to suite an intermediate standard release, and the few available test handsets are not implemented to work at this standard. In 1999 ETSI finished the standardisation for UMTS Release 1999. In March 2001 UMTS Release 4 specification was frozen and in March 2002 UMTS Release 5 (the initial target date was December 2001). The release 4 will be compatible with any further releases, and can thus be seen as a viable basis to build networks on.

Licensing and rollout

The European countries have attributed licences differently, on the basis of some basic requirements concerning the technical standard and basic procedures for attribution (NMTC 2001). The basic regulation directives from EU sat a time frame for an UMTS authorisation system by January 1st 2000, the primary condition for introduction of UMTS services by January 1st 2002. At least one license had to be attributed in each country, and the number of licences could only be limited by limitations in available frequencies. Licenses had to be granted on open, non-discriminatory and transparent procedures, based on objective criteria defined in advance.

The attribution procedures in Europe vary from auctions, beauty contests (comparative selections) or a mixture of the two. The final licence costs for the licensees vary according to the procedures, as the auctions resulted in relatively large incomes for the attributing countries, and accordingly high expenses for the licensees. The high start up fees imposed by auctions is argued to be positive, because it will ensure that the licensees must utilize the technology at its very best. The possibility that the users must pay the price is at the same time present, but is considered to be avoided by effective competition and the sunk cost in infrastructure. On the other hand, the beauty contests favors the established actors that in a contest with few applicants only have to give modest obligations, or on the contrary create a competition where the obligation of coverage and speed of development offered is more than the applicants economically can achieve. Year based frequency fees also vary, as for example Norway implementing a fee on 20 million NOK per licence while France operates with 1 per cent of the turnover. The endurance1 of the licences also varies from 12 year in Norway, to United Kingdom with 20 years (with a possible extension of another 20 years). The obligations concerning the time frames of rollout is also varying, the United Kingdom on the extreme where obligations for rollout is due not before 2007. The issues concerning operators sharing the networks are also dealt with differently in the different countries and the regulation policy are varying in their degree of specificity. The fragmented conditions this attribution process has created may result in highly different markets, as well as influencing the decisions of pan-European operators concerning entry of a new national market.

The case of Norway

The introduction of UMTS in Norway can be seen as a result of the imposed EU obligations. At the same time, the Scandinavian countries nature of being up front when it comes to telecommunication has indeed played an important role. Anyway, the main responsibility for the success of UMTS in Norway, by delivery of services that the public want to use at prices they can afford, lies with the different stakeholders as operators, equipment manufacturers as well as service providers.

Because of acknowledged uncertainties concerning the market for UMTS as well as the technology, the Norwegian parliament (The Storting) decided to let the applicants, that supposedly could do better assessments of the potential demand for services and cost of infrastructure development, to compete on the basis of planed coverage and speed of rollout, as a beauty contest. The relatively modest up front cost of licences was supposed to be seen together with the obligation to develop and rollout the networks as a partial payment of the licences (NMTC 2001). In Norway, the demand for coverage, equal prices and quality in the entire country was seen as the superior target that best would be achieved by beauty contest in comparison with auctions.

The Norwegian Ministry of Transport and Communications gave an invitation to tender for UMTS licences 31st of May 2000 (NMTC 2000b). Within the deadline 14th of August the same year, seven applications was received (NMTC 2000a), and by first of December four applicants where granted licences: *Tele2 Norge AS, Telenor ASA, NetCom GSM AS and Broadband Mobile ASA*. The award process and the conditions of evaluation were decided upon by the Storting (NMTC 1999). The applicants had to fulfil the minimum requirements of extent of coverage, speed of rollout and financing. In addition the licences were granted on an evaluation of additional geographical coverage and increased rollout speed offered by the applicants. The applications were also evaluated on the basis of financial aspects, quality of telephony services, commitment to environmental concerns, experiences in similar operations and further development of the Norwegian mobile telephony market (NMTC 2000a). The most offensive licensee in the beauty contest; Tele2, initially planned to within an eight year period reach 4.393.000 inhabitants of Norway, and cover a geographical area of 271.956 square kilometres by building 4.283 base-stations (NMTC 2001).

The licences both came with rights and obligations. The licensees were granted the permission to exclusively use specified frequencies in a twelve-year period from when the license was awarded. The licensees had to pay a start-up fee of 200 million NOK and a fee each year (currently 20 million NOK) to use the frequencies. In a five year period after the licence are awarded, a minimum of 90 per cent of homes in the 12 most densely populated areas of Norway must be covered, and each operator's network must appear as a separate network to avoid reduced competition because of cooperation. This implies that no roaming between UMTS networks nationally is allowed. At the same time, the operators also have the right and duty to co-locate with other providers when it comes to base-stations and antennas.

The state of the game

None of the Norwegian licensees has made to follow their licence obligations in the first year of development (2001). Tele2 Norge AS, the most offensive licensee in the beauty contest, has not deployed *any* of the 821 base-stations self-obligated in the beauty contest for implementation in 2001, but only signed contracts with landowners on the rights to build on a total of 444 sites (NPTA 2002a). These delays are explained by uncertainties and lack of availability of network infrastructure equipment and software. The equipment available has also been based on an UMTS version that not will be commercial deployed. At the same time, no commercial handsets have been developed for the intermediate standard, and handsets for the frozen standard is delayed, forecasted to be available in late 2002. Still, the delays of Tele2 can only to a small extent be explained by factors out of their control, as the work of the other station points require neither specifications nor standards (NPTA 2002a). The delays should therefore also be understood as a result of strategic decisions.

During 2001 the global financial environment faced a slowdown, and telecommunication was no exception to this, rather one of the most suffering sectors. There was a sharp increase in debt level, mainly by very high UMTS-licence costs. The availability of investments funds also became significantly reduced, creating a major problem for new entrants in the market. This came when the rollout of infrastructure and the preparation of services required significant economical resources. To keep up with this situation operators have changed their main focus from UMTS rollout to increase ARPU (average revenue per user) on existing technology (2 and 2.5G). A high priority has also been given to reduce investments costs of UMTS by influencing regulation bodies to allow a higher degree of network infrastructure sharing. In this environment the producers of equipment faces the pressure from operators on prices and the reduced sales of GSM equipment because of market saturation and operators' reduced investment in subsidising terminals. UMTS handsets are still not available, and the dual-mode terminals (backwards compatible with 2G networks) are still at design and early testing stages facing known problems concerning: Dropped calls, glitches in the terminal software, and insufficient battery capacity.

At the same time, which key applications and services these handsets need to serve is also still an open question. As more than 50 operators in Europe are launching their networks within a relatively short time frame, bottlenecks in equipment delivery obviously could occur when the rollout finally happens.

The financial situation, among other factors has resulted in some EU countries without sufficient number of interested applicants for licences, early rollout obligations has not been met due to availability of equipment, and only experimental and not commercial networks are in operation. Authorities have met this situation differently, as Italy has increased licence duration from 15 to 20 years. Portugal and Spain have already decided to postpone the early obligation of development and rollout. In France, the initial fee of 4.95 Billion NOK was reduced to 619 Million NOK per licence after the awards.

Still, no services are available as there are no UMTS networks in operation. And without a range of services UMTS will probably provide no extra revenue to the operators and will not be attractive to end-user compared with GSM. As the services are still to come, the handsets not in the market and the network not rolled out and in operation, the European context conceals five vicious circles inhibiting the development of the UMTS market (EU 2002a):

- Industry faces difficulties in providing affordable handsets suiting the content services, due to higher cost of data-enabled handsets as well as operators unable to subsidise. Expensive handsets will result in lower penetration, further impacts on operator revenues, in turn decreasing the operators ability to subsidise the handsets
- The operator's cost of licenses disables them to add enabling features in billing system upgrades and user localisation systems. This does not stimulate development of innovative content services, again worsening the economical situation of the operators
- The financial situation of the operators also reduces their willingness to provide access to their platforms to third parties, as well as imposing revenue sharing that is unbalanced from the content

provider point of view. This does not stimulate for growth of the content market
- The limited availability of proper handsets creates a "wait and see" attitude from the service providers
- Proliferation of standards for the packaging and transmission of content over mobile infrastructure requires spread over multiple technical options, again justifying delay of investment from content providers.

The challenges to face

To meet revised and more moderate expectations of UMTS in Europe today compared to those existing few years ago, three processes must be accomplished:
- Rollout of networks and operation of them
- Make and provide suitable terminals
- Provide attractive applications, content and services.

Until now, all of these processes have failed to show success. Further delays and problems in one of these processes will create further distortions and prohibit the success of UMTS, as described in the vicious circles. The nature of these delays has recently been reported as the shipment of handsets from the major manufacturer Nokia no longer are due in September 2002, but early 2003 depending on the operator schedules of the networks (Nokia 2002). On the other hand, the not so offensive any longer licensee Tele2 have not only failed to meet its initial obligations, but also the revised obligations to implement a test network of 20 base-stations in urban Oslo within the 15th of July 2002. Due to this, the Ministry of Transport and Communications have issued day-based fines from 15th of July currently 200.000 NOK, rising to 400.000 NOK after 14 days (NPTA 2002b).

EU (EU 2002) describes the consensus for the work ahead in the current situation: Let the market operate and the rolling out of UMTS is an evolving process. Further, four critical layers which may influence these processes and the final deployment and success of UMTS in EU (EU 2001):
- The regulatory environment
- The financial context
- Gaining experience with the new market
- Outstanding technical issues.

The regulatory environment should be harmonised and stabilised across the EU. Currently, operators have to face 15 different regulatory regimes in the European markets. The regulation bodies should also only take action where the market fails to create correct pricing of the services as well as market development, and changes in rollout obligations should only occur if on the basis of transparent and objective conditions. Changes in the licence duration are not seen as of major concern, but changes in fees and charges are not recommended, indeed if increased. At the same time, clarification and harmonising of the rules for network infrastructure sharing (NIS) will ease the situation for the operators. Lately, there have been suggestions that infrastructure sharing, and thus cost sharing, should be allowed to avoid economical problems for the licensees, but then at the cost of competition. As described, the different EU countries have taken measures to ease the situation of the licensees, but by different means, adding to the heterogeneity in the European market.

The financial context is clearly depicted with the economical burden of the licences, and at least the cost of deploying and marketing the networks. This burden must be seen in relation with the optimistic situation in the economy at the point of granting the licences, both reflected in the high bids in the auctions as well as the high burdens of rollout taken on where beauty contests where used. The economical downturn has also resulted in a significantly decreased commercial value of the UMTS spectrum. These financial aspects will have implications on the ability for newcomers to enter the market, as well as the front-end investment might negatively effect the development of UMTS services.

The experience with the market is missing, although the use of innovative 2G services accessible with WAP (Wireless Application Protocol) have proven useful feedback, but not fulfilled all expectations. The main uncertainties concerns business models and revenue sharing among operators, service providers and content providers, as well as which services the operators will offer and what prices the users will accept. The question whether the end-users will pay the cost of the licenses as well as the new transmission infrastructure is still open, and must be seen together with the users evaluation of the services provided by UMTS.

Licensees are postponing the acquisition of network infrastructure equipment, as they probably must be replaced in the new future, as well as the prices probably will drop. Until this these uncertainties are handled, as well as until the handsets compatible with the standard are accessible, there will probably be a wait-and-see situation. The basic cellular network infrastructure middleware such as wireless application protocol (WAP) and micro-web-browser for the handsets is also not agreed upon, and different handset producers are also using varying operation systems on their UMTS handsets.

UMTS as innovation in telecommunication

Innovations can be defined as: "... the introduction and spread of new and improved products and processes in the economy." (Freeman et al 1997 p24). This definition emphasis the differences between innovation and invention, the latter the novel ideas, models and plans that *might* become an innovation. Godø follows this definition, and identifies innovation as inventions that are brought into wide spread use and become economical successful (Godø 1995). Godø also emphasise the nature of being "new to the world" as an important feature of inventions and thus innovations.

Innovations will appear different from different perspectives, and although it must be something "new to the world"; it is not required to be new for the end-users. Godø exemplifies this by describing the introduction of the maritime INMARSAT satellite communication system that was a novel innovation for R&D in telecommunication in terms of the satellite technology, but to the end-users, nothing more than improved quality in use was gained compared with the short wave radio used prior to satellite (Godø 1995 p72). The novelty of innovation will also depend on the level of analysis: On the basis of components, innovations can be "new to the world", and at the same time considered only as a part of an existing system. Innovation can also be introduced by combining existing components in a novel arrangement.

Innovations in telecommunications are identified both to be radical and incremental by Godø (Godø 1995). The radical innovations are not attributed to serendipity, chance or other haphazard processes, but a result of usually large, often with global cooperation, R&D efforts for a certain purpose (Godø 1995). Still purpose doesn't include the necessity of the knowledge of or the mere existence of end-user needs, as Godø describes telefax (Godø 1995 p44-45), and Haug describes the introduction of first generation of mobile telecommunication system in Scandinavia. Both these innovation were technological viable from the R&D perspective, but from the end-user perspective, innovations (at least radical) in telecommunications don't come on demand, and users usually have no idea what they want since the medium is a new one to most of them. This is one important factor, which makes assessment of whether the innovation will break its critical mass of users: "... impossible to predict *ex ante – ex post* explanations are usually difficult to falsify.". Technological push, as this strategy of innovation may be called, is thus not only risky but also a necessity, as these markets must be created.

By introducing innovation regimes in telecommunication, Godø gives an explanation of the nature of innovation in the telecommunication sector. With a focus on the role of R&D in the process of innovation, he identifies the existence of two regimes of innovation in telecommunication: An innovation driven and a diffusion driven. The innovation driven regime is by Godø coined Large Technological Projects (LTPs), and dominated by telecommunication R&D, national and internationally, delivering radical system innovations. This regime consists of a framework of institutions as well as international cooperation based on long traditions, making it possible to radically introduce new systems by technological push on a large scale. In Norway, Telenor (former "Televerket") have historically worked close with the Nordic and the European telecommunication operators. Through this cooperation, not only significant resources for R&D investments becomes available, but also the competence and talent, technically as politically, enabling the creation of radical innovation possible. The diffusion driven regime or Large Technological Networks (LTNs) does on the other hand exist much closer to an existing "market", where innovations supports incremental system growth as well as exploring network externalities. LTNs only foster incremental innovations, as the installed base of an existing system only can bear minor adjustments as telecommunication systems are based on interoperability. The range of freedom for innovations is restricted, and innovations follow conservative technological trajectories to explore network externalities.

UMTS and stakeholders

As noted earlier, and acknowledged by EU, the "... 3G rollout is by fare more complex than the deployment of the second generation (2G) of mobile communications." (EU 2002b). The complexity of the current innovation process of 3G is partly a result of the increasing number of stakeholders as well as the complex interaction between different actors.

The original key driver for UMTS in Europe was the standardisation bodies ITU and ETSI, representing the national telecommunication operators. Their initial work was in line with the approach followed by first and second generation of mobile communication systems, and thus rarely disturbed in their work with their standardisation efforts. Their aim was to introduce a new mobile telecommunication system that could be used globally, as well as utilising the frequency spectrum more efficient. The consequences of the political decisions done by these bodies may have important effects on national R&D efforts, as illuminated when ITU in 1982 voted down the frequency allocation for a maritime mobile communication system suggested by R&D in "Televerket", resulting in the project to close down (Godø 1995 p77-78). This also illustrates the range of frequency administration decision: Because the system was designed to operate on a global scale, reallocation of frequencies had to be done globally.

Godø (Godø 1995) identifies operators, equipment manufacturers and end-users as the most important actors in the dynamics of technological development in telecommunications. As time has passed by, actors as national and international governmental bodies, as well as content service providers have introduced their influence in the innovation and diffusion process of UMTS. As the standardisation bodies have created a common ground and consensus for the UMTS system, the further development and rollout is to a great extent out of their hands.

OPERATORS

Initially, through their participation in standardisation bodies, the state-owned operators ("Televerket") had full control in the development of standards, innovation and rollout. Today, the R&D departments of the operators are still the sources of innovations, but as the markets are liberalised they are facing competition both nationally, and from pan-European operators. Thus, the liberalisation of the telecommunication market have resulted in a situation where monopoly has been replaced with competition, both on market shares but also when it comes to control over the innovation and rollout process as a whole concerning UMTS.

Traditionally the operators have not focused on competition and have been active in creation of national telecommunication industry, as well as securing national goals of telecommunication accessibility and use. The development of 1G experienced (*suffered* in Haugs terminology) very little interference from outside forces, both from manufacturers, users and governmental bodies (Haug 2002). 2G had a different experience, when more operators were represented as well as interferences from political and industrial interest as well as EU became apparent, as EU early expressed the need to speed up the development of GSM. Operators no longer have the role, or the interest of securing national strategies, and national as well as international governmental bodies no longer delegate these responsibilities directly, but rather through regulation. This has resulted in a more open operator market, but also in the introduction of regulation mechanisms to secure effective competition creating substantial uncertainty for the operators.

As "Televerket" become Telenor, and the Norwegian telecommunication market was deregulated and liberalised, other operators have been introduced in Norway. But, beyond doubt and clearly depicted by market shares and number of employees, Telenor still are in control of the Norwegian telecommunication market, as well as expanding in other countries as Ukraine, Malaysia, Bangladesh, Portugal etc. Even if they are concerned and being cautious not to violate competition legislations, they both have the financial resources as well as the R&D experience and competence to be the major source of innovations in the Norwegian context.

As services no longer are solely voice or simple data transfer, services are supposed to be supplied by third parties. Without these services, UMTS will not appear as an innovation for the end-users. On the other hand, operators also rely on the providers of handsets. As the numbers of handset provider increase and the operators economical ability to subsidise is low due to high licence cost, as well as handsets are shipped to be used not by any single operator but in Europe in general, operators have no ability to dictate the specifications of the handsets. In the current situation of the implementation of UMTS, operators are not in full control. Still, they have a strategic position when it comes to develop and rollout networks, but without deployment of handsets and services, the network is worthless. It is obvious that strategic decisions done by operators favours a wait and see attitude by the other actors, as well as possible resulting in introduction of UMTS as a part of the GSM infrastructure with a few new services.

EQUIPMENT MANUFACTURERS

The equipment manufacturers deliver both the network infrastructure and the handsets. The specification of network infrastructure has been frozen, and the technology is currently commercial delivered by for example Nokia. This equipment has been developed as equipment to the former generations of mobile communications, based on formal specifications from the operators. The time of deliverance of this equipment has of course played an important role when it comes to timing of the rollout of UMTS, and delays in this process have been used as one of the explanation of the delays in the whole of the rollout process.

The handset manufacturers not only have to conform to the specification of the operators and the network, but also the new services. The development of handsets faces difficulties, as the doubt concerning the existence of operational networks, as well as the nature of services still is missing. The manufacturers are also not shipping handsets if networks are not in operation. On the other hand, handsets supporting certain services might create *de facto* standards in this early phase, and thus result in favourable market shares.

Historically, manufacturers of handsets have worked close with the operators, and have followed strict specifications. More autonomy has been achieved especially by Nokia implementing 2.5G services enabling communication of animations and ring-tones in the network, but still within the specification of the networks of the operators. UMTS opens up a range of new possibilities, and the nature of the handsets, having their own operating systems etc., gives great freedom for the specifications of the handsets. On the other hand it has created a high degree of uncertainty as the services still are to be defined, as well as a range of user-needs must be created. Together with technical challenges created by the need for a smart design, larger colour screens as well as increased battery capacity, the uncertainty of services as well as the lack of network infrastructure has resulted in not available handsets.

END-USERS

End-users are also thought to be a driver, and of course will be the critical mass in it self for UMTS as the success of UMTS is built on the prognosis of end-user needs for mobile multimedia services. It is rather obvious that the introduction of UMTS will have to follow the line of its predecessors by a process of technological push. By pushing the technology, eventually a critical mass will be created as the users discover their needs, or as these needs are created. In this situation the end-users plays a minor, if any role at all. But with a wider perspective, the end-user as currently 2G (and 2.5G) users, play an important role in their increasing demand and use of these services showing a soaring market opportunity in telecommunication in general. But this also makes it convenient for the operators to stick to these existing and revenue producing technologies that actually may face competition from UMTS. As services still are not defined, it is hard assess whether the end-user market will be as expected. As the development of UMTS has faced difficulties, questions concerning actual UMTS coverage and the economical context of third party service development are important factors for the end-users. In addition, the strong position of Telenor, as well as the obligation of the operators to build independent networks might result in expensive services, as well as services only accessible in urban areas. As UMTS most probably will be introduced as dual handsets for GSM and UMTS, 2G (and 2.5G) will probably be the common services in early use of the UMTS networks. But this is not enough for the operators that needs increased use and traffic to cover the expenses of the new network, requiring the use of multimedia services.

GOVERNMENTAL BODIES

EU and its member states aim to create a telecommunication market with effective competition. The networks are not only implemented to reach national and international goals of accessibility and quality, but also as competence and technology that can be exported worldwide when the networks are in operation. UMTS is also assumed as a primary enabling factor strengthening the ability of European to compete globally in any sector, based on high penetration and use of information technology.

The national and international governmental bodies responsible for granting operating licences make decisions that have major effects on telecommunication projects and the environment for rollout of the networks. This is both positive, as the power of EU makes decisions possible, but can also have negative consequences as decision may have (possible negative) effects on markets.

The strict policy-prohibiting network sharing among the operators exemplifies this. By allocation of frequencies the introduction of UMTS has been enabled globally, but introducing licence fees and rollout obligations slow the introduction process down. The acts of governments have also resulted in uncertainties, as the licence agreements might be changed anytime to suite the current development of the markets, as well as the heterogeneity of regimes in the different European countries.

In the case of Norway, the attribution process used can be argued to have failed to create a sound market. This is due to the bankruptcy of one of the licences and the seemingly overambitious obligations given by Tele2. The bankruptcy can be explained as a result of other economical circumstances, but the strategy of Tele2 should be explained by the problems associated with beauty contests: Some applicants give obligations far more ambitious than applicable to achieve a licence. This argument is further strengthened by the fact that both Telenor and NetCom, which was obvious candidates for licences, only gave modest obligations. This does however not automatically imply that auctions would have created a better market situation.

SERVICE PROVIDERS

The service providers are supposed to drive the diffusion and the further rollout of UMTS networks by introducing high-valued services to end-users. Developed by third parties, value-adding services will also be the main source of revenue produced by UMTS compared to its predecessors. In the current situation there are no if any services, naturally caused by the lack of operational and available networks, as well as handsets.

A central question concerning the UMTS services is the revenue sharing models. As service providers usually have to build services on the basis of content providers, operators will no longer be alone to collect the revenue. In Norway, the operators have traditionally showed little interest of sharing the revenue equally or in a sustainable manner with the service providers. This of course doesn't foster development of new services. Probably, this will change in the future, and revenue sharing will into a greater extent favour service providers creating traffic in the UMTS networks. Even tough, the initial services will probably be developed and provided by the operators them selves. These services may at the same time only be available for the subscriber of one single operator.

UMTS as an innovation

In the case of UMTS presented here, the distinction between invention and innovation is obvious if we compare the existing specification and ideas of 3G with the non-existence of commercial and available UMTS-networks for the end-user. Still, the borders between the invention of UMTS and its forecasted precedence of an innovations, becomes blurry when we emphasise different part or level of the "product", as well as which market we emphasise. It has also become obvious that the transition from the invention of UMTS to the innovation and implementation of operational UMTS networks has become an incremental and will be an ongoing process.

The new UMTS-handsets or devices can be seen as an invention itself, as they must provide large colour screens, higher bandwidth capacity, more memory and processing power, thus "new to the world". To support this, the handsets also need higher battery capacity, potentially requiring further inventions. At the same time, they have to be compatible with the existing GSM networks, and probably only be used as ordinarily GSM-phones until a range of UMTS services is present. The new base-stations also require inventions to support UMTS, as well as the content services itself must be suited and optimised to the new handsets. In sum, UMTS implementations are not based on a single invention, but consist of a range of inventions originating from different sources, which together will make up a successful innovation. The single of these inventions will not become widespread (more than for test purposes) and have economical impact without the successful introduction of the others.

For the end-user, UMTS will become an innovation when they find the services provided worthwhile using. This will of course be reflected by the service and content providers that only will be able to collect their share of the revenue when services finally are used. The network operators as licensees have already acknowledged UMTS as an (coming) innovation by significant investments in licenses, as well as with their obligations to rollout infrastructure. The auctions, and the beauty contest in the case of Norway have already have economical impact for the licensees, both when it comes to licence fees and the cost of rollout where rollout has started. Accordingly the countries awarding licenses have already had significant income, and (hopefully) secured diffusion of technology by attaching rollout and diffusion obligations to the licensees. On the other hand, the rather non-existing service providers have still not invested much in development of new and value adding services, and several of the potential service providers have also gone bankrupt since 2000 as they have failed to survive providing services to 2G (and 2.5G).

As a consequence of liberalization of the communication market as well as the introduction of multiple new actors, the control structures and enabling character of a LTP regime are not present when it comes to total control over the further invention and rollout of UMTS. It is rather obvious that UMTS prevails in the regime of LTNs in its further development, diffusion and innovation. Still, the initial work with UMTS in Norway reflects a quite different strategy.

In a historical context, it is natural to se the idea of UMTS conceived in the LTP regime. In the mid eighties, the majority of European telecommunication operators where state owned, and the radical innovations and rollout of NMT and GSM were supposed to be preceded by yet another success. As noted, the development and diffusion of these technologies where done almost autonomously, and without interference from governmental agencies.

At the same time, at least concerning NMT, the numbers of stakeholders was limited, and consensus making was possible within reasonable time frames. In this context, without the existence of markets and user needs, technology push was not only a possibility but also a necessity. But as time has passed by, the nature of the context has changed as the number of stakeholders has increased dramatically. The influence of governmental bodies are high as they find the future development of the national and regional development of the telecommunication sector as strategically vital, as well as cooperation among the operators have turned into competition. In sum, the current state of the game in 2002 is revealing a situation of a LTN regime.

Not only have the process of creating the UMTS standard showed that consensus making is getting more time consuming, but the presence of equipment manufacturers and third party service providers shows that R&D in telecommunication is not any longer in control of the operators but the mechanisms of the market. At the same time, the strategic decisions of licensees and handset manufacturers, as well as the absence of services and third party service providers, underscore a wait and see policy where each of the stakeholders desperately tries to exploit as much as possible from 2 and 2.5G.

The introduction and diffusion of UMTS is showing a process where no single actor has the possibility to introduce the innovation as such. This is new to the Norwegian telecommunication sector. EU acknowledges the major challenge of these dependencies when it comes to broadband infrastructure in general: "Infrastructure investment is driven by availability of content and services and the development of new content and services depends on infrastructure deployment." (EU 2002c p8). Breaking out from the vicious circles described above, EU is determined to create a virtuous circle where infrastructure and services together will exploit network externalities. To achieve this EU suggests an actions plan where policy measures will strengthen competition and interoperability, where good practise is disseminated, where the process will be monitored by benchmarking and finally where overall co-ordination of existing policies brings synergies (EU 2002c).

This shows that EU still have confidence in the ongoing diffusion process of UMTS, but also leaving the real actions to the market it self. At the same time this market is shaped by a strong regulation regime favouring the established actors, and in particular Telenor. This may lead to an effective market in the future, but as for now; there is still a way to go before UMTS become an innovation for the end-users in Norway.

6

Bluetooth

OVERVIEW

Bluetooth is an always-on, short-range radio hook-up that resides on a microchip. It was initially developed by Swedish mobile-phone maker Ericsson in 1994 as a way to let laptop computers make calls over a mobile phone. Since then, several thousand companies have signed on to make Bluetooth the low-power short-range wireless standard for a wide range of devices.

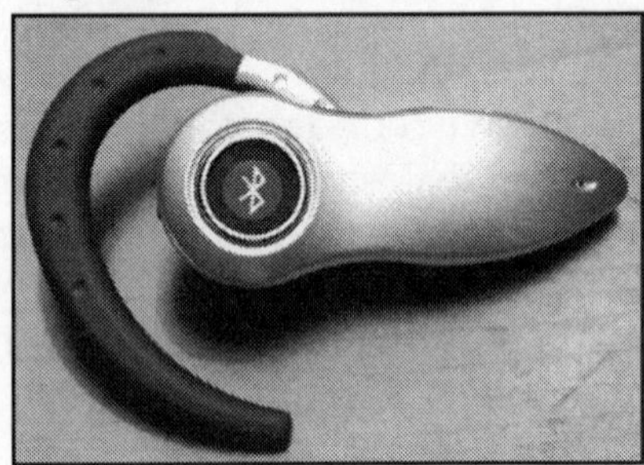

The Bluetooth standards are published by an industry consortium known as the Bluetooth SIG (special interest group).

The concept behind Bluetooth is to provide a universal short-range wireless capability. It uses the 2.4-GHz band, available globally for unlicensed low-power uses. Bluetooth is intended to support an open-ended list of applications including data, audio, graphics, and even video. Bluetooth can provide consumers with the ability to do the following:

- Make calls from a wireless headset connected remotely to a cell phone.
- Eliminate cable linking computers to printers, keyboards, and the mouse.
- Hook up MP3 players wirelessly to other machines to download music.
- Set up home networks so that normal household equipment such as air conditioning, lightning, oven controls etc. can be remotely monitored and controlled.

- Call home from a remote location to turn appliances on or off, set alarms and monitor activity.

BLUETOOTH VS. WI-FI (IEEE 802.11)

Bluetooth and Wi-Fi (the brand name for products using IEEE 802.11 standards) have some similar applications: setting up networks, printing, or transferring files. Wi-Fi is intended as a replacement for cabling for general local area network access in work areas. This category of applications is sometimes called wireless local area networks(WLAN). Bluetooth was intended for portable equipment and its applications. The category of applications is outlined as the wireless personal area network (WPAN). Bluetooth is a replacement for cabling in a variety of personally carried applications in any setting and also works for fixed location applications such as smart energy functionality in the home (thermostats, etc.).

Wi-Fi is a wireless version of a common wired Ethernet network, and requires configuration to set up shared resources, transmit files, and to set up audio links (for example, headsets and hands-free devices). Wi-Fi uses the same radio frequencies as Bluetooth, but with higher power, resulting in higher bit rates and better range from the base station. The nearest equivalents in Bluetooth are the DUN profile, which allows devices to act as modem interfaces, and the PAN profile, which allows for ad-hoc networking.

DEVICES

Fig. A Bluetooth USB dongle with a 100 m range.

Bluetooth exists in many products, such as telephones, tablets, media players, Lego Mindstorms NXT, PlayStation 3, PS Vita, the Nintendo Wii, and some high definition headsets, modems, and watches. The technology is useful when transferring information between two or more devices that are near each other in low-bandwidth situations. Bluetooth is commonly used to transfer sound data with telephones (i.e., with a Bluetooth headset) or byte data with hand-held computers (transferring files). Bluetooth protocols simplify the discovery and setup of services between devices. Bluetooth devices can advertise all of the services they provide. This makes using services easier because more of the security, network address and permission configuration can be automated than with many other network types.

COMPUTER REQUIREMENTS

Fig. A typical Bluetooth USB dongle.

Fig. An internal notebook Bluetooth card (14×36×4 mm).

A personal computer that does not have embedded Bluetooth can be used with a Bluetooth adapter that will enable the PC to communicate with other Bluetooth devices (such as mobile phones, mouse and keyboards). While some desktop computersand most recent laptops come with a built-in Bluetooth radio, others will require an external one in the form of a dongle.

Unlike its predecessor, IrDA, which requires a separate adapter for each device, Bluetooth allows multiple devices to communicate with a computer over a single adapter.

Operating system implementation

Apple products have worked with Bluetooth since Mac OS X v10.2 which was released in 2002. For Microsoft platforms, Windows XP Service Pack 2 and SP3 releases work natively with Bluetooth 1.1, 2.0 and 2.0+EDR.[28]Previous versions required users to install their Bluetooth adapter's own drivers, which were not directly supported by Microsoft. Microsoft's own Bluetooth dongles (packaged with their Bluetooth computer devices) have no external drivers and thus require at least Windows XP Service Pack 2. Windows Vista RTM/SP1 with the Feature Pack for Wireless or Windows Vista SP2 work with Bluetooth 2.1+EDR. Windows 7 works with Bluetooth 2.1+EDR and Extended Inquiry Response (EIR).

The Windows XP and Windows Vista/Windows 7 Bluetooth stacks support the following Bluetooth profiles natively: PAN, SPP, DUN, HID, HCRP. The Windows XP stack can be replaced by a third party stack which may support more profiles or newer versions of Bluetooth. The Windows Vista/Windows 7 Bluetooth stack supports vendor-supplied additional profiles without requiring the Microsoft stack to be replaced.

Linux has two popular Bluetooth stacks, BlueZ and Affix. The BlueZ stack is included with most Linux kernels and was originally developed by Qualcomm. The Affix stack was developed by Nokia. FreeBSD features Bluetooth since its 5.0 release. NetBSD features Bluetooth since its 4.0 release. Its Bluetooth stack has been ported to OpenBSD as well.

MOBILE PHONE REQUIREMENTS

A Bluetooth-enabled mobile phone is able to pair with many devices. To ensure the broadest feature functionality together with compatibility with legacy devices, the Open Mobile Terminal Platform (OMTP) forum has published a recommendations paper, entitled "Bluetooth Local Connectivity".

SPECIFICATIONS AND FEATURES

The Bluetooth specification was developed as a cable replacement in 1994 by Jaap Haartsen and Sven Mattisson, who were working for Ericsson in Lund, Sweden. The specification is based on frequency-hopping spread spectrum technology. The specifications were formalized by the Bluetooth Special Interest Group (SIG). The SIG was formally announced on 20 May 1998. Today it has a membership of over 17,000 companies worldwide. It was established by Ericsson, IBM, Intel, Toshiba and Nokia, and later joined by many other companies. All versions of the Bluetooth standards are designed for downward compatibility. That lets the latest standard cover all older versions.

Bluetooth v1.0 and v1.0B

Versions 1.0 and 1.0B had many problems, and manufacturers had difficulty making their products interoperable. Versions 1.0 and 1.0B also included mandatory Bluetooth hardware device address (BD_ADDR) transmission in the Connecting process (rendering anonymity impossible at the protocol level), which was a major setback for certain services planned for use in Bluetooth environments.

Bluetooth v1.1

- Ratified as IEEE Standard 802.15.1–2002
- Many errors found in the 1.0B specifications were fixed.
- Added possibility of non-encrypted channels.
- Received Signal Strength Indicator (RSSI).

Bluetooth v1.2

Major enhancements include the following:

- Faster Connection and Discovery

- *Adaptive frequency-hopping spread spectrum (AFH),* which improves resistance to radio frequency interference by avoiding the use of crowded frequencies in the hopping sequence.
- Higher transmission speeds in practice, up to 721 kbit/s, than in v1.1.
- Extended Synchronous Connections (eSCO), which improve voice quality of audio links by allowing retransmissions of corrupted packets, and may optionally increase audio latency to provide better concurrent data transfer.
- Host Controller Interface (HCI) operation with three-wire UART.
- Ratified as IEEE Standard 802.15.1–2005
- Introduced Flow Control and Retransmission Modes for L2CAP.

Bluetooth v2.0 + EDR

This version of the Bluetooth Core Specification was released in 2004. The main difference is the introduction of an Enhanced Data Rate (EDR) for faster data transfer. The nominal rate of EDR is about 3 Mbit/s, although the practical data transfer rate is 2.1 Mbit/s. EDR uses a combination of GFSK and Phase Shift Keying modulation (PSK) with two variants, ð/4-DQPSK and 8DPSK. EDR can provide a lower power consumption through a reduced duty cycle.

The specification is published as "Bluetooth v2.0 + EDR" which implies that EDR is an optional feature. Aside from EDR, there are other minor improvements to the 2.0 specification, and products may claim compliance to "Bluetooth v2.0" without supporting the higher data rate. At least one commercial device states "Bluetooth v2.0 without EDR" on its data sheet.

Bluetooth v2.1 + EDR

Bluetooth Core Specification Version 2.1 + EDR was adopted by the Bluetooth SIG on 26 July 2007.

The headline feature of 2.1 is secure simple pairing (SSP): this improves the pairing experience for Bluetooth devices, while increasing the use and strength of security. See the section on Pairing below for more details.

2.1 allows various other improvements, including "Extended inquiry response" (EIR), which provides more information during the inquiry procedure to allow better filtering of devices before connection; and sniff subrating, which reduces the power consumption in low-power mode.

Bluetooth v3.0 + HS

Version 3.0 + HS of the Bluetooth Core Specification was adopted by the Bluetooth SIG on 21 April 2009. Bluetooth 3.0+HS provides theoretical data transfer speeds of up to 24 Mbit/s, though not over the Bluetooth link itself. Instead, the Bluetooth link is used for negotiation and establishment, and the high data rate traffic is carried over a collocated 802.11 link.

The main new feature is AMP (Alternative MAC/PHY), the addition of 802.11 as a high speed transport. The High-Speed part of the specification is not mandatory, and hence only devices sporting the "+HS" will actually support the Bluetooth over 802.11 high-speed data transfer. A Bluetooth 3.0 device without the "+HS" suffix will not support High Speed, and needs to only support a feature introduced in Core Specification Version 3.0 or earlier Core Specification Addendum 1.

L2CAP Enhanced modes Enhanced Retransmission Mode (ERTM) implements reliable L2CAP channel, while Streaming Mode (SM) implements unreliable channel with no retransmission or flow control. Introduced in Core Specification Addendum 1.

Alternative MAC/PHY Enables the use of alternative MAC and PHYs for transporting Bluetooth profile data. The Bluetooth radio is still used for device discovery, initial connection and profile configuration, however when large quantities of data need to be sent, the high speed alternative MAC PHY 802.11 (typically associated with Wi-Fi) will be used to transport the data. This means that the proven low power connection models of Bluetooth are used when the system is idle, and the faster radio is used when large quantities of data need to be sent. AMP links require enhanced L2CAP modes.

Unicast Connectionless Data Permits service data to be sent without establishing an explicit L2CAP channel. It is intended for use by applications that require low latency between user action and reconnection/transmission of data. This is only appropriate for small amounts of data.

Enhanced Power Control Updates the power control feature to remove the open loop power control, and also to clarify ambiguities in power control introduced by the new modulation schemes added for EDR. Enhanced power control removes the ambiguities by specifying the behaviour that is expected. The feature also adds closed loop power control, meaning RSSI filtering can start as the response is received. Additionally, a "go straight to maximum power" request has been introduced. This is expected to deal with the headset link loss issue typically observed when a user puts their phone into a pocket on the opposite side to the headset.

Ultra-wideband

The high speed (AMP) feature of Bluetooth v3.0 was originally intended for UWB, but the WiMedia Alliance, the body responsible for the flavor of UWB intended for Bluetooth, announced in March 2009 that it was disbanding, and ultimately UWB was omitted from the Core v3.0 specification.

On 16 March 2009, the WiMedia Alliance announced it was entering into technology transfer agreements for the WiMedia Ultra-wideband (UWB) specifications. WiMedia has transferred all current and future specifications, including work on future high speed and power optimized implementations, to the Bluetooth Special Interest Group (SIG),Wireless USB Promoter Group and the USB Implementers Forum. After the successful completion of the technology transfer, marketing and related administrative items, the WiMedia Alliance will cease operations.

In October 2009 the Bluetooth Special Interest Group suspended development of UWB as part of the alternative MAC/PHY, Bluetooth v3.0 + HS solution. A small, but significant, number of former WiMedia members had not and would not sign up to the necessary agreements for the IP transfer. The Bluetooth SIG is now in the process of evaluating other options for its longer term roadmap.

Bluetooth v4.0

The Bluetooth SIG completed the Bluetooth Core Specification version 4.0 and has been adopted as of 30 June 2010. It includes *Classic Bluetooth, Bluetooth high speed* and*Bluetooth low energy* protocols. Bluetooth high speed is based on Wi-Fi, and Classic Bluetooth consists of legacy Bluetooth protocols.

Bluetooth low energy (BLE), previously known as WiBree, is a subset of Bluetooth v4.0 with an entirely new protocol stack for rapid build-up of simple links. As an alternative to the Bluetooth standard protocols that were introduced in Bluetooth v1.0 to v3.0, it is aimed at very low power applications running off a coin cell. Chip designs allow for two types of implementation, dual-mode, single-mode and enhanced past versions. The provisional names *Wibree* and *Bluetooth ULP* (Ultra Low Power) were abandoned and the BLE name was used for a while. In late 2011, new logos "Bluetooth Smart Ready" for hosts and "Bluetooth Smart" for sensors were introduced as the general-public face of BLE.

- In a single mode implementation the low energy protocol stack is implemented solely. CSR, Nordic Semiconductor and Texas Instruments have released single mode Bluetooth low energy solutions.
- In a dual-mode implementation, Bluetooth low energy functionality is integrated into an existing Classic Bluetooth controller. Currently (2011-03) the following semiconductor companies have announced the availability of chips meeting the standard: Qualcomm-Atheros, CSR, Broadcom and Texas Instruments. The compliant architecture shares all of Classic Bluetooth's existing radio and functionality resulting in a negligible cost increase compared to Classic Bluetooth.

Cost-reduced single-mode chips, which enable highly integrated and compact devices, feature a lightweight Link Layer providing ultra-low power idle mode operation, simple device discovery, and reliable point-to-multipoint data transfer with advanced power-save and secure encrypted connections at the lowest possible cost.

General improvements in version 4.0 include the changes necessary to facilitate BLE modes, as well the Generic Attribute Profile (GATT) and Security

Manager (SM) services with AES Encryption. Core Specification Addendum 2 was unveiled in December 2011; it contains improvements to the audio Host Controller Interface and to the High Speed (802.11) Protocol Adaptation Layer.

BLUETOOTH APPLICATIONS

Bluetooth is designed to operate in an environment of many users. Up to eight devices can communicate in a small network called a piconet. Ten of these piconets can coexist in the same coverage range of the Bluetooth radio.

To provide security, each link is encoded and protected against eavesdropping and interference.

Using short range wireless connectivity, Bluetooth can provide support for three general applications areas:

- Data and voice access point: Bluetooth facilitates real-time voice and data transmissions by providing wireless connection of portable stationary communications devices.
- Cable Replacement: Connections are instant and are maintained even when the devices are not within LoS. The range of each radio is approximately 10 m but can be extended to 100 m with an optional amplifier.
- Ad hoc networking: A device equipped with a Bluetooth radio can establish instant connection to another radio as soon as it comes into range.

Other Bluetooth Application examples are:

- Three in one phone: Phone works as an intercom (office), fixed line (home), or mobile (when on the move).
- Internet Bridge: Use a portable PC to surf in the internet anywhere, whether the connection is wireless through a mobile phone (mobile) or through a wired connection (PSTN, ISDN, LAN, xDSL).
- Briefcase e-mail: Notification of received e-mail message using the mobile phone.
- Automatic Synchronization: of desktop computer, portable PC, notebook and mobile phone.
- Cordless Desktop: Connect your desktop/laptop computer cordlessly to printers, scanners, keyboard, mouse, and the LAN.

BLUETOOTH STANDARDS DOCUMENTS

Bluetooth standards represent a formidable bulk. There are well over 1500 pages, divided into two groups:

- Core Specifications: Describes the details of the various layers of the Bluetooth protocol architecture, from the radio interface to link control.
- Profile Specifications: Are concerned with the use of Bluetooth technology to support various applications. Each profile specification

discusses the use of the technology defined in the core specifications to implement a particular usage model. It includes a description of which aspects of the core specifications are mandatory, optional, and not applicable.

Protocol Architecture

Bluetooth is defined as a layered protocol architecture consisting of core protocols, cable replacement, telephony control protocols, and adopted protocols .

The core protocols form a five-layer stack consisting of the following elements:

- *Radio:* Specifies the details of the air interface, including frequency, the use of frequency hopping, modulation scheme, and transmit power.
- *Baseband:* Concerned with connection establishment within a piconet, addressing, packet format, timing, and power control.
- *Link Manager Protocol (LMP):* Responsible for linking setup between Bluetooth devices and ongoing link management. (this includes security aspects such as authentication and encryption, plus the control and negotiation of baseband packet sized).
- *Logical Link Control and Adaptation Protocol (L2CAP):* Adapts upper-layer protocols to the baseband layer. L2CAP provides both connectionless and connection-oriented services.
- *Service Discovery Protocol (SDP):* Device information, services, and the characteristics of the services can be queried to enable the establishment of a connection between two or more Bluetooth devices.

RFCOMM is the cable replacement protocol included in the Bluetooth specification. RFCOMM presents a virtual serial port that is designed to make a replacement of cable technologies as transparent as possible.

RFCOMM provides for binary data transport and emulates EIA-232 control signals over the Bluetooth baseband layer. EIA-232 (formerly known as RS-232) is a widely used serial port interface standard.

Bluetooth specifies a telephony control protocol. TCS BIN (telephony control specification-binary) is a bit oriented protocol that defines the call control signalling for the establishment of speech and data calls between Bluetooth devices.

The adopted protocols are defined in specifications issued by other standards-making organizations and incorporated into the overall Bluetooth architecture. The Bluetooth strategy is to invent only necessary protocols and use existing standards whenever possible. Some of the adopted protocols are:

- PPP (point to point protocol)
- TCP/UDP/IP

- OBEX (object exchange protocol)
- WAE/WAP (wireless application environment and protocol).

Usage Models

In essence, a usage model is a set of protocols that implement a particular Bluetooth-based application. Each profile defines the protocol and protocol features supporting a particular usage model. The highest priority usage models are:

- File Transfer: Supports the transfer of directories, files, documents, images, and streaming formats. This usage model also includes the capability to browse folders on a remote device.
- Internet Bridge: With this usage model, a PC is wirelessly connected to a mobile phone or cordless modem to provide dial-up networking and fax capabilities.
- LAN access: This usage model enables devices on a piconet to access a LAN.
- Synchronization: This model provides a device-to-device synchronization of personal information management (PIM) information.
- Three in one phone: Telephone handsets that implement this usage model may act as a cordless phone connecting to a voice BS, as an intercom device for connecting to other telephones, and as a cellular phone.
- Headset: The headset can act as a remote device's audio input and output interface.

Piconets and Scatternets

The basic unit of networking in Bluetooth is a piconet, consisting of a master and from one to seven active slave devices. The radio designated as the master makes the determination of the channel (Frequency Hopping Sequence) and phase (timing offset, i.e., when to transmit) that shall be used by all devices on this piconet.

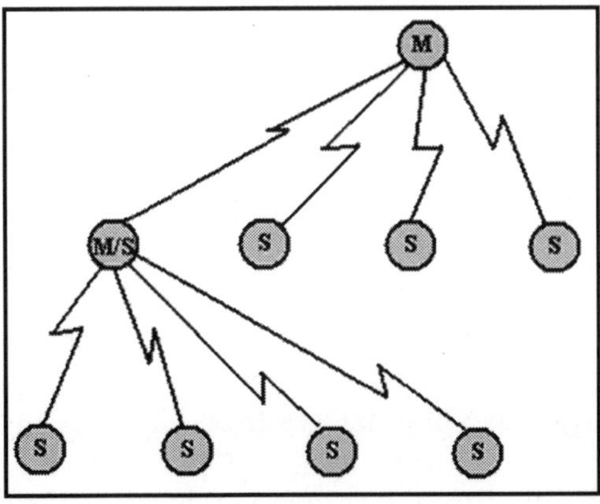

Fig. Master/Slave Relationship

The radio designated as master makes this determination using its own device address as a parameter, while the slave devices must tune to the same channel and phase. A slave may only communicate with the master and may only communicate when granted permission by the master. A device in one piconet may also exist as a part of another piconet and may function as either a slave or master in each piconet as illustrated in Figure. This form of overlapping is called a scatternet. The piconet/scatternet architecture in comparison with other forms of wireless networks. The advantage of the piconet/scatternet scheme is that it allows many devices to share the same physical are and make efficient use of the bandwidth. The physical area and total bandwidth are shared by the scatternet. The logical channel and data transfer are shared by a piconet.

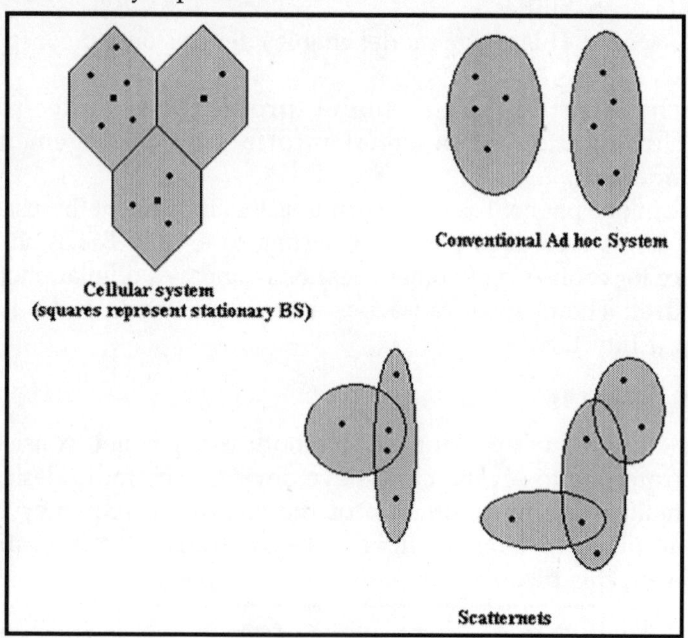

Fig. Wireless Network Configurations.

Radio Specifications

The Bluetooth radio specification is a document that gives the basic details of radio transmission for Bluetooth devices. Table below summarizes the key parameters:

Bluetooth makes use of the 2.4-GHz band within the ISM band. In most countries, the bandwidth is sufficient to define 79 1-MHz physical channels. Power control is used to keep the devices from emitting any more RF power than necessary.

The power control algorithm is implemented using the link management protocol between a master and the slaves in a piconet.

Table. Bluetooth Radio and Baseband Parameters.

Parameter	Value
Topology	Up to 7 simultaneous links in a logical star
Modulation	GFSK
Peak data rate	1 Mbps
RF bandwidth	220 kHz (-3dB), 1 MHz (-20dB)
RF band	2.4-GHz ISM band
RF carriers	23/79
Carrier spacing	1 MHz
Transmit power	0.1 W
Piconet access	FH-TDD-TDMA
Frequency hop rate	1600 hops/s
Scatternet access	FH-CDMA

One important aspect of the radio specifications is the definition of three classes of transmitters based on output power:

- Class 1: Outputs 100 mW (+20 dBm) for maximum range, with a minimum of 1 mW (0 dBm). In this class power control is mandatory, ranging from 4 to 20 dBm. This mode provides the greatest distance.
- Class 2: Outputs 2.4 mW (+4 dBm) at maximum, with a minimum of 250 mW (-6 dBm). Powr control is optional.
- Class 3: Lowest power. Nominal output is 1 mW.

Modulation for Bluetooth is Gaussian FSK, with a binary one represented by a positive frequency deviation and a binary zero represented by a negative frequency deviation from the centre frequency. The minimum deviation is 115 kHz.

The main characteristics of the receivers include the following:

- Sensitivity Level: The receiver must have a sensitivity level for which the bit error rate (BER) 0.1 per cent is met. For Bluetooth this means an actual sensitivity level of -70dBm or better.
- Interference Performance: The interference performance on Co-channel and adjacent 1 MHz and 2 MHz are measured with the wanted signal 10 dB over the reference sensitivity level. On all other frequencies the wanted signal shall be 3 dB over the reference sensitivity level.
- Out-of-Band blocking: The Out of band blocking is measured with the wanted signal 3 dB over the reference sensitivity level. The interfering signal shall be a continuous wave signal. The BER shall be less than or equal to 0.1 per cent.
- Intermodulation Characteristics: The reference sensitivity performance, BER = 0.1 per cent, shall be met under the following conditions.
- The wanted signal at frequency f_0 with a power level 6 dB over the reference sensitivity level.

- A static sine wave signal at f_1 with a power level of –39 dBm.
- A Bluetooth modulated signal at f_2 with a power level of -39 dBm.

Such that:

$$f_0 = 2 \cdot f_1 - f_2 \quad \text{and}$$
$$|f_2 - f_1| = n \cdot 1\,\text{MHz},$$

where n can be 3, 4, or 5. The system must fulfil one of the three alternatives.

- *RSSI: Receiver Signal Strength Indicator (Optional):* A transceiver that wishes to take part in a power-controlled link must be able to measure its own receiver signal strength and determine if the transmitter on the other side of the link should increase or decrease its output power level. A Receiver Signal Strength Indicator (RSSI) makes this possible.

The way the power control is specified is to have a golden receive power range. This golden receive power is defined as a range with a lower and higher threshold levels and a high limit. The lower threshold level corresponds to a received power between -56 dBm and 6 dB above the actual sensitivity of the receiver. The upper threshold level is 20 dB above the lower threshold level to reach an accuracy of ±6 dB. The instructions to alter the TX power are carried in the LMP link

Baseband Specifications

The Baseband is the physical layer of the Bluetooth. It manages physical channels and links apart from other services like error correction, data whitening, hop selection and Bluetooth security.

The Baseband layer lies on top of the Bluetooth radio layer in the Bluetooth stack. The baseband protocol is implemented as a Link Controller, which works with the link manager for carrying out link level routines like link connection and power control.

The baseband also manages asynchronous and synchronous links, handles packets and does paging and inquiry to access and inquire Bluetooth devices in the area. The baseband transceiver applies a time-division duplex (TDD) scheme. (alternate transmit and receive). Therefore apart from different hopping frequency (frequency division), the time is also slotted.

Frequency Hopping Frequency hopping in Bluetooth serves two purposes:

- Provides resistance to interference and multipath effects
- Provides a form of multiple access among co-located devices in different piconets.

The total bandwidth is divided into 79 physical channels, each of bandwidth 1 MHz. FH occurs by jumping from one channel to another in a pseudorandom sequence. The hopping sequence (PN) is shared by all of the devices on a single piconet.

The hop rate is 1600 hops per second, so that each physical channel is occupied for a duration of 625-ms. Each of the 625-ms time period is referred to as a slot, and these are numbered sequentially.

Bluetooth radios communicate using a time division duplex discipline (TDD). Because more than two devices share the piconet medium, the access technique is TDMA. Thus, piconet access can be characterized as FH-TDD-TDMA.

Using TDD prevents crosstalk between the transmit and receive operations in the radio transceiver, which is essential if a one-chip implementation is desired.

The FH sequence is determined by the master in a piconet and is a function of the master's Bluetooth address. Because different piconets in the same area will have different masters, they will use different hop sequences. Thus, most of the time, transmission on two devices on different piconets in the same area will be on different physical channels.

Occasionally, two piconets will use the same physical channel during the same time slot, causing a collision and lost data. However, because this will happen infrequently, it is readily accommodated with forward error correction and error detection/ARQ techniques. Thus, a form of CDMA is achieved between devices on different piconets in the same scatternet; this is referred to as FH-CDMA.

Physical Links

Two types of links can be established between a master and a slave:

- Synchronous connection oriented (SCO): Allocates a fixed bandwidth between a point-to-point connection involving the master and a single slave. The master maintains the SCO link by using reversed slots at regular intervals. The basic unit of reservation is two consecutive slots (one in each transmission direction). The master can support up to three simultaneous SCO links while a slave can support two or three SCO links. SCO packets are never retransmitted.
- Asynchronous connectionless (ACL): A point-to-multipoint link between the master and all the slaves in the piconet. In slots not reserved for SCO links, the master can exchange packets with any slave on a per-slot basis, including a slave already engaged in an SCO link. Only a single ACL link can exists. For most ACL packets, packet retransmission is applied.

Packets

The packet format for all Bluetooth packets consists of three main fields:

- Access Code (72 bits): Used for timing synchronization, offset compensation, paging and inquiry.

- Header (54 bits): Used to identify packet type and to carry protocol control information.
- Payload (0 to 2745 bits): If present, contains user voice and data and, in most cases, a payload header.

There are three types of access code:

- Channel access code (CAC): Identifies the piconet (unique for a piconet).
- Device access code (DAC): Used for paging and its subsequent responses
- Inquiry access code(IAC): Used for inquiry purposes.

The header format for all Bluetooth packets consists of six fields:

- AM_ADDR(3 bits): Contains the "active mode" address of one of the slaves (there are at most seven devices in a piconet). A transmission from the master to a slave contains that slave's address; a transmission from a slave contains it address. The 0 value is reserved for a broadcast from the master to all slaves in the piconet.
- Type: Identifies the type of packed. Four type codes are reserved for control packets common to both SCO and ACL links. The remaining packet types are used to convey user information.
- Flow: Provides a 1-bit flow control mechanism for ACL traffic only. When a packet with Flow = 0 is received, the station receiving the packet must temporarily halt the transmission of ACL packets on this link. When a packet with Flow = 1 is received, transmission may resume.
- ARQN: Provides 1-bit acknowledgement mechanism for ACL traffic protected by a CRC. If the reception was successful, an ACK (ARQN = 1) is returned; otherwise a NAK (ARQN = 0) is returned. When no return message regarding acknowledge is received, a NAK is assumed implicitly. If a NAK is received, the relevant packet is retransmitted.
- Header error control (HEC): An 8-bit error detection code used to protect the packet header.

Payload Format

For some packet types, the baseband specification defines a format for the payload field.

For voice payloads, no header is defined. For all of the ACL packets and for the data portion of the SCO packet, a header is defined. For data payloads, the payload format consists of three fields:

- Payload header: An 8-bit header is defined for a single-slot packet, and a 16-bit header is defined for a multi-slot packet.

- Payload body: Contains user information
- CRC: A 16-bit CRC code is used on all data payloads except the AUX1 packet (which carries 30 information bytes with no CRC or FEC which is typically used for high-speed data.)

The payload header, when present consists of three fields:

- L_CH: Identifies the logical channel.
- Flow: Used to control flow at the L2CAP level. (This is the same on/ off mechanism provided by the flow field in the packet header for ACL traffic).
- Length: The number of bytes of data in the payload, excluding the payload header and CRC.

Error Correction

Bluetooth makes use of three error correction schemes:

- 1/3 rate FEC
- 2/3 rate FEC
- ARQ

These error correction schemes are designed to satisfy competing requirements. The error correction scheme must be adequate to cope with the inherently unreliable wireless link but must also be streamlined and efficient.

Logical Channels

Bluetooth defines five types of logical data channels designed to carry different types of payload traffic.

- Link Control (LC): Used to manage the flow of packets over the link interface. The LC channel is mapped onto the packet header. This channel carries low-level link control information like ARQ, flow control, and payload characterization.
- Link Manager (LM): Transports link management information between participating stations.
- User asynchronous (UA): Carries asynchronous user data.
- User isochronous (UI): Carries isochronous user data.
- User synchronous (US): Carries synchronous user data.

Channel Control

The operation of a piconet can be understood in terms of the states of operation during link establishment and maintenance. There are two major states:

- Standby: The default state. This is a low-power state in which only the native clock is running.
- Connection: The device is connected to a piconet as a master or a slave.

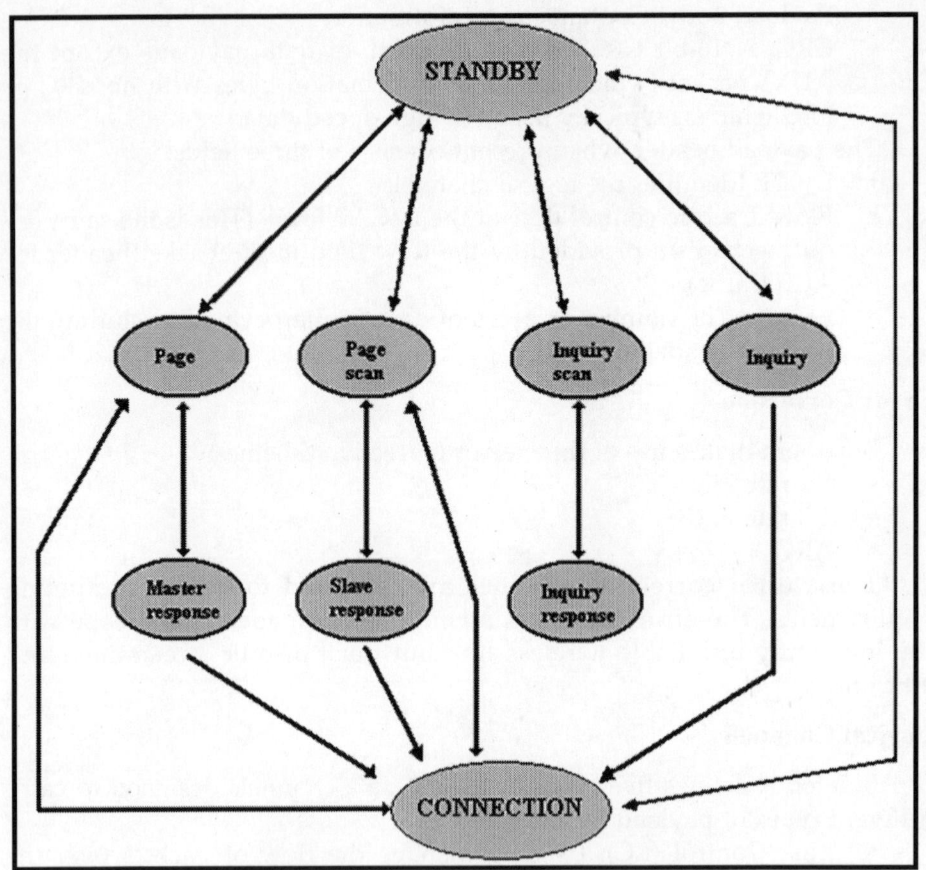

Fig. Bluetooth state transition diagram.

In addition, there are seven interim sub-states that are used to add new slaves to a piconet.

To move from one state to the other, either commands from the Bluetooth link manager or internal signals in the link controller are used. The sub-states are as follows:

- *Page:* Used by the master to activate and connect to a slave. Uses the slave device access code (DAC)
- *Page scan:* Device is listening for a page with its own DAC.
- *Master response:* A device acting as a master receives a page response from a slave. The device can now enter the connection state or return to the page state to page for other slaves.
- *Slave response:* A device acting as a slave responds to a page from a master. If connection setup succeeds, the device enters the connection state; otherwise it returns to the page scan state.
- *Inquiry:* Device has issued an inquiry, to find the identity of the devices within range.

- *Inquiry scan:* Device is listening for an enquiry
- *Inquiry response:* A device that has issued an inquiry receives an inquiry response.

BLUETOOTH AUDIO

The baseband specifications indicate that either of two voice encoding schemes can be used:

- Pulse Code Modulation (PCM)
- Continuously Variable Slope Delta (CVSD)

BLUETOOTH SECURITY

The Bluetooth baseband specifications defines a facility for link security between any two Bluetooth devices, consisting of the following elements:

- Authentication.
- Encryption (Privacy).
- Key management and usage.

The security algorithms make use of four parameters:

- Unit address: The 48-bit device address, which is publicly known.
- Secret authentication key: A secret 128-bit key
- Secret privacy key: A secret key of length from 4 to 128 bits
- Random number: A 128-bit random number derived from a pseudorandom generation algorithm executed in the Bluetooth unit.

The two secret keys are generated and configured with the unit and not disclosed.

The purpose of authentication is to verify the claimed identity of one of the two Bluetooth devices involved in an exchange. User information can be protected by encryption of the packet payload.

Link Manager Specifications

The Link Manager carries out link setup, authentication, link configuration and other protocols. It discovers other remote LM's and communicates with them via the Link Manager Protocol (LMP). To perform its service provider role, the LM uses the services of the underlying Link Controller (LC).

The Link Manager Protocol essentially consists of a number of PDU (protocol Data Units), which are sent from one device to another. LM PDUs are always sent as single-slot packets and the payload header is therefore one byte.

The procedures defined for LMP are grouped into 24 functional areas, each of which involves the exchange of one or more messages. Lists these areas, together with the PDUs involved in each area.

Table. LMP PDU's.

Function	PDU's
General response	Accepted, not accepted
	Security Service
Authentication	au_rand, sres
Pairing	in_rand, au_rand, sres, comb_key, unit_key
Change link key	comb_key
Change current link key	temp_rand, temp_key, use_semi_permanent_key
	Time/synchronization
Clock offset request	clkoffset_req, clkoffset_res
Slot offset information	slot_offset
Timing accuracy Information request	timing_accuracy_req, timing_accuracy_res
	Station Capability
LMP version	version_req, version_res
Supported features	features_req, features_res
	Mode Control
Switch master/slave role	switch_req
Name request	name_req, name_res
Detach	detach
Hold mode	hold, hold_req
Sniff mode	sniff, sniff_req, sniff_res
Park mode	park_req, park_set, set_broadcast_window, modify_beacon, unpark_PM_ADDR_req, unpark_BD_ADDR_req
Power control	incr_power_req, decr_power_req, max_power, min_power
Channel quality-driven Change between DM and DH	auto_rate, preferred_rate
Quality of service	quality_of_service, quality_of_service_req
SCO links	SCO_link_req, remove_ SCO_link_req
Control of multi-slot packet	max_slot, max_slot_req
Paging scheme	page_mode_req, page_scan_mode_req
Link supervision	supervision_timeout.

Logical Link Control and Adaptation Protocol (L2CAP)

In the IEEE 802 specification, L2CAP provides a link layer protocol between entities across a shared-medium network. It also provides a number of services and relies on the baseband layer for flow and error control.

L2CAP makes use of ACL links; it does not provide support for SCO links. L2CAP provides 2 alternatives of service to upper-layer protocols using ACL links:

- Connectionless service: This is a reliable datagram style of service.

- Connection-mode service: This is similar to the service offered by HDLC. A logical connection is set up between two users exchanging data. Flow and error control are provided.

L2CAP provides three types of logical channels:

- Connectionless: Supports the connectionless service. Each channel is unidirectional and is used for broadcast from the master to multiple slaves.
- Connection-oriented: Supports connection oriented service. Each channel is full duplex. A quality of service flow specification is assigned in each direction.
- Signalling: Provides fro the exchange of signalling messages between L2CAP entities.

L2CAP Packets

Figure below shows the format of L2CAP packets:

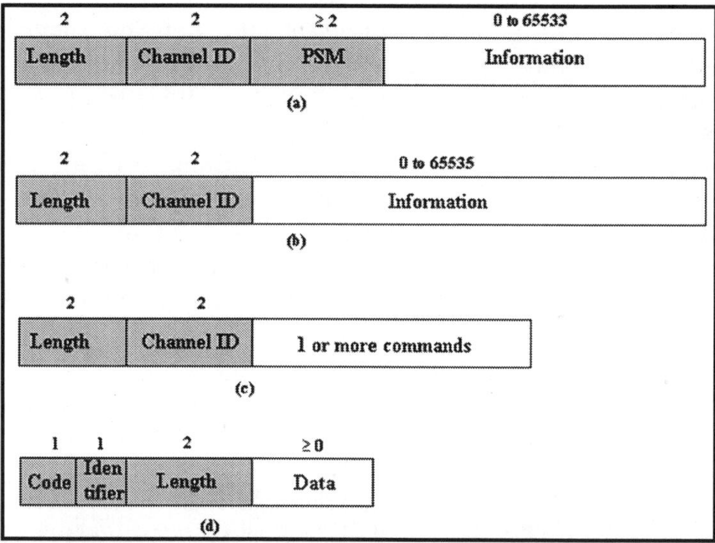

Fig. L2CAP Packet Format. (a) Connectionless PDU. (b) Connection-oriented PDU. (c) Signalling command PDU. (d) Command Format.

In the connectionless PDU, the following fields are used:

- Length: Length of the information payload plus PSM fields in bytes
- Channel ID: A value of 2 indicating the connectionless channel.
- Protocol/Service multiplexer (PSM): Identifies the higher layer recipient for the payload in this packet.
- Information payload: Higher layer user data. This field may be up to 65533 bytes in length.

Connection oriented packets have the same format as connectionless packets, but without the PSM field. The PSM field is not needed because the CID identifies the upper layer recipient of the data. The information payload field may be up to 65535 bytes is length. Signalling commands packets have the same header format as the connection-oriented packets. In this case the CID value is 1, indicating the signalling channel. The payload of a signalling packet consists of one or more L2CAP commands, each of which consists of four fields:

- Code: Identifies the type of command
- Identifier: Used to match a request with its reply. The requesting device sets this field and the responding device uses the same value in its response. A different identifier must be used for each original command.
- Length: Length of the data field for this command in bytes
- Data: Additional data, if necessary, relating to this command.

Signalling Commands

There are eleven commands in five categories. The command reject command can be sent in response to any command to reject it. Reasons for rejection include invalid CID or length exceeded.

- Connection Commands: Used to establish a new logical connection.
- Configure Commands: Are sent to establish an initial logical link transmission contract between two L2CAP entities and to renegotiate this contract whenever appropriate
- Disconnection Commands: Are used to terminate a logical channel
- Echo commands: Are used to solicit a response from a remote L2CAP entity. These commands are typically used for testing the link or passing vendor-specific information using the optional data field.
- Information Commands: Are used to solicit implementation-specific information from a remote L2CAP entity.

Table. L2CAP signalling command codes.

Code	Description	Parameters
0x01	Command reject	Reason
0x02	Connection request	PSM, Source CID
0x03	Connection response	Destination CID, Source CID, Result, Status
0x04	Configure request	Destination CID, Flags, Options
0x05	Configure response	Source CID, Flags, Results, Options
0x06	Disconnection request	Destination CID, Source CID
0x07	Disconnection response	Destination CID, Source CID
0x08	Echo request	Data (optional)
0x09	Echo response	Data (optional)
0x0A	Information request	Info Type
0x0B	Information response	Info Type, Result, Data (optional)

Quality of Service (QoS)

The QoS parameter in L2CAP defines a traffic flow specification based on RFC 1363. In essence, a flow specification is a set of parameters that indicate a performance level that the transmitter will attempt to achieve.

The flow specification consists of the following parameter:

- Service type: This parameter indicates the level of service for this flow. A value of 0 indicates that no traffic will be transmitted on this channel. A value of 1 indicates a best effort service. The device will transmit data as quick as possible with no guarantees about performance. A value of 2 indicates a guaranteed service (The sender will transmit data that conform to the remaining QoS parameters).
- Token rate (bytes/second): See below (bucket size)
- Token bucket size (bytes): This parameter, along with the Token rate parameter defines a token bucket scheme that is often used in QoS specifications. The advantage of this scheme is that it provides a concise description of the peak average traffic load the recipient can expect.
- Peak bandwidth (bytes/second): Limits how fast packets may be sent back-to-back from applications.
- Latency (microseconds): Is the maximum acceptable delay between transmission of a bit by the sender and its initial transmission over the air.
- Delay variation (microseconds): is the difference between the maximum and minimum possible delay that a packet will experience.

7

Guidance for Sustainable Mobile Telecommunication Growth

Telecommunications is defined as 'all forms of communications by electrical or optical wire and cable and radio signals (whether terrestrial or from satellite), both public and private' (PPG 8).

Domestic satellite dishes will not be dealt with in this SPG. Government guidance is provided in the Department of the Environment, Transport and the Regions (DETR) publication 'A Householder's Planning Guide for the Installation of Satellite Television Dishes' (Copies available from the Borough Planning Office at the Civic Centre). This guidance is itself currently under review by the Office of the Deputy Prime Minister (ODPM).

Growth in consumer demand for mobile telecommunications and operational requirements to improve the quality and range of services has raised a number of public concerns in a Borough which is crossed by major transport routes, has significant areas of Green Belt and Metropolitan Open Land, an historic built environment and extensive residential areas.

The Government considers that modern telecommunications systems have a vital part to play in peoples' lives, bringing significant economic and social benefits. In recent years the rapid expansion of the sector has made telecommunications a major component of the national economy. However there have been concerns as to the siting, appearance and health issues relating to telecommunication installations.

By appreciating the essential infrastructure requirements of the network providers and the potential impact on the environment, Hounslow Council seeks to ensure by working with network providers and interested parties that telecommunication developments are sensitively delivered within the Borough.

This Supplementary Planning Guidance aims to elaborate:

- Hounslow's telecommunications policy
- Provide guidance for telecommunications operators
- Specify Hounslow's siting and appearance requirements for telecommunication installations.

Planning Policy Guidance Note 12 (Development Plans) states that:

- SPG may be taken into account as a material consideration in the determination of planning applications. The Secretary of State will give substantial weight to an SPG which has evolved from the development plan, and has been prepared in the proper manner.
- A public consultation should form part of the SPG preparation and include interested parties. These views should be taken into account before the SPG is finalised.

MOBILE TELECOMMUNICATIONS: TELECOM TECHNOLOGY EVOLUTION

In order to fully understand the future of mobility, it is important to examine the evolution of individual streams of mobile telecommunication technology. These developments have not happened independently of each other. As mobility-enabling technology improves, three streams of telecommunication technology—networks, devices, and applications—will become truly smart in the future. Networks will converge, devices will become wearable, and context-aware applications will act spontaneously and initiate actions.

In the past, enterprise adoption was a key driver of technology commoditization. By contrast, recent progress in mobile technology has been largely shaped by consumer demand. Enterprises need vision to see how mobility will change over time so that they can help drive its evolution and bring about economies of scale.

A NEW ERA OF MOBILE TELECOMMUNICATIONS

Telecom technology is changing our lives in ways we did not envision. Today, efficient packet switched networks carry media packets (voice, data, and video) to intelligent mobile devices that users carry everywhere.

The past 50 years have witnessed a migration from traditional telephony networks to telephony networks based on an IP backbone and the cloud. Today's telecom technology can be effectively divided into three streams—networks, devices, and applications. Following the evolution of each stream provides background and context for understanding what the future holds.

The Evolution of Networks

The evolution of the Telecom Service Provider (TSP) wireless networks. We see that consumer demand shaped the evolution of wireless networks.

The Evolution of Devices

Mobile device technology has undergone changes, and its evolution has been largely consumer driven. With the increasing availability of low-cost bandwidth, mobile devices are used not just for communication and content creation, but also for web browsing with rich graphics, and for tasks such as handling workflows. The first driver of device technology evolution is mobile operating systems. Enterprise customers saw the dramatic rise in popularity of the Blackberry OS, particularly for email access. The future could include decoupling of devices and operating systems, where end-users will effectively be able to run the OS of their choice on any device as long as the hardware is compatible.

Second, input device technology for the user interface (UI) has changed as well. Devices have moved from the keypad-based interfaces on the older mobile phones to touchscreens. Touchscreens are also undergoing evolution as they provide an increasing number of features such as multi-touch gestures. Today, in addition to touch-based input devices, gaming consoles use gesture-based remotes. In the future, haptic gestures may be captured by sensors on mobile devices, and the interface could be a projection onto a blank screen.

System on a Chip (SoC) integrates a processor, a Graphics Processing Unit (GPU), and memory controllers on a single chip. This integration improves overall system performance. With the advent of quad core processors, complex 3D display systems are possible. Nanotechnology is promising in this area.

Enterprise-Driven Technology Evolution

Mobility technology evolution has traditionally been consumer-driven rather than enterprise-driven. This slow and steady adoption of new technologies by enterprises stems from the balance they seek to create between the benefits that each new technology brings and its maturity at the time of implementation. The inherent need for stability and risk aversion has meant that enterprise customers adopt technologies only after they mature and are deemed stable?

At the same time, the enterprise has the potential to catapult any technology into mainstream by providing a scale of consumption and related efficiencies that are necessary to ensure mass adoption of new communications technologies.

From an adoption perspective, we see a shift from the traditional pattern of enterprise-enterprise-individual to an individual-enterprise-individual pattern. Improved connectivity and mobility in users' personal lives has heightened connection expectations of employees, customers, shareholders, field workers, and other stakeholders. In fact, the point of inflection has already been reached, beyond which the cumulative sales of smartphones and tablets will exceed the number of PCs, laptops, and netbooks sold globally. It will also mark the first time that any such inflection point has been driven by the consumer as opposed to the enterprise.

The future promises more consumer segment driven changes, bolstering the case for enterprises to accelerate the pace at which they adopt and adapt to new technologies. Resisting change brings the risk of being unable to capitalize on the productivity that the new era of mobility and the significant productivity gains that each subsequent transition will bring.

Enterprise customers need to move faster and ride new technology transitions with less adoption latency while embracing the new communication modes and models that result from these transitions into their business processes. Those that do so will effectively be creating highly evolved, self-optimizing business models and environments that are capable of continually discovering new sources of highly profitable revenue growth. As the future unfolds, evolution in telecommunications will help create a digital utopia for both consumers and enterprises, where everyone is connected, and smart devices become an extension of our consciousness.

HISTORY OF MOBILE TELECOMMUNICATIONS

The Mobile Telecommunications (MT) group, headed by Prof. Di Yuan, was founded in 2006. The group's research develops new concepts, models, algorithms, and protocols in design, analysis, and performance engineering of wireless communication networks.

The Internet is going wireless. The key enabler of this development is the rapid deployment of high speed mobile access networks, ranging from 3G and beyond-3G cellular, WiMAX, mesh, to the integration of these technologies. There are two key performance aspects in this emergence. The first is the evolution of the physical-layer transmission technology, and the second is performance engineering in a networking perspective. The latter is the core of the research excellence of the MT group.

Our research covers a wide spectrum of performance engineering aspects in wireless communications. Classical topics in the field are radio resource management, routing in multi-hop radio networks and traffic engineering, and cross-layer optimization. Real-time bandwidth measurement, quality of service assurance, as well as mobility management form major research themes of the group.

A key expertise area is network optimization that enables a radio access network to maximally exploit the capacity of the transmission layer. Optimization applies to the deployment of radio elements, its parameter configuration, and resource allocation. Optimization is also a powerful tool in understanding the performance limits of wireless communications.

The group's research extends to the applications of mobile communication technologies in other major sectors of the society. A good example is wireless networking for intelligent transportation systems.

Taking an integrated approach that has its roots in complementary scientific tools, including mathematical optimization, queuing theory, algorithmic analysis, and computer modeling and simulation, the MT group's research delivers cutting edge methodology to performance engineering in the era of wireless Internet.

For additional information of the group, and its research and teaching activities, please navigate to the respective pages using the menu to the left.

News

2013

- *Sep. 2013:* The FP7-ICT-SMARTCITIES project RERUM is launched.
- *Aug. 2013:* Professor Xingjun Zhang from the Xi'an Jiaotong University, China, is seconded to MT for 3 months.
- *Jun. 2013:* Scott Fowler joins the Editorial Board of the *International Journal of Embedded Systems* (IJES) as Associate Editor.
- *May 2013:* Vangelis Angelakis joins the Editorial Board of the *Journal of Commmunications and Networks* (JCN).
- *May 2013:* Scott Fowler is seconded, within an EC-FP7 Marie Curie IRSES project, to the Xi'an Jiaotong University in China, for 5 months.
- *May 2013:* Vangelis Angelakis and Qing He are seconded, within an EC-FP7 Marie Curie IRSES project, to the Huazhong University of Science and Technology (HUST) in Wuhan, China, for 5 and 3 months respectively.
- *May 2013:* David Gundlegård and Vangelis Angelakis co-author of the 19 papers in the Mobile/Transport session of the D4D Challenge at NetMob2013 at MIT.
- *Apr. 2012:* The MT group receives one EC-FP7 Marie Curie IAPP grant award, with invitation from the European Research Executive Agency (REA) for grant negotiation.
- *Apr. 2013:* Lei Chen defends his PhD thesis.
- *Mar. 2013:* The FP7-ICT-SMARTCITIES proposal RERUM rankes 2nd out 43 submissions. Vangelis Angelakis, Scientist in Charge for MT, is invited for negotiations with the EC.
- *Mar. 2013:* EC-FP7 Marie Curie Project MESH-WISE, coordinated by the MT group (Di Yuan), is officially launched.
- *Jan. 2013:* David Gundlegård presents his Licentiate Thesis.

2012

- *Dec. 2012:* Di Yuan receives a grant award from the Swedish Foundation for Strategic Research (SSF) for research with Ericsson in 2013.
- *Dec. 2012:* Di Yuan gives lecturers at an ELLIIT PhD course in Lund University, Sweden.
- *Nov. 2012:* Di Yuan receives an EC-FP7 Marie Curie Career award, with invitation from the European Research Executive Agency (REA) for grant negotiation.
- *Nov. 2012:* By invitation of Institute for Infocomm Research (A-Star, Singapore), Di Yuan gives presentation at the Advanced Wireless Communications Workshop.
- *Aug. 2012:* Di Yuan and the MT group receive two EC-FP7 Marie

Curie grant awards, with invitation from the European Research Executive Agency (REA) for grant negotiation.

- *Jun. 2012:* Iana Siomina (Ericsson, formerly an MT member) and Di Yuan win a Best Paper Award at the IEEE International Communications Conference (ICC) 2012. ICC is the flagship conference of the IEEE Communications Society (ComSoc).
- *Jun. 2012:* Joyce Wu joins MT as guest researcher within an EC-FP7 Marie Curie IAPP project.
- *Apr. 2012:* Wei Jing joins MT as guest researcher within an EC-FP7 Marie Curie IAPP project.
- *Mar. 2012:* Di Yuan starts an exchange activity at Ranplan (UK) within an EC-FP7 funded Marie Curie IAPP project.
- *Feb. 2012:* Di Yuan joins the Editorial Board of the *Computer Networks* journal (Elsevier) as Area Editor.
- *Feb. 2012:* Di Yuan visits the University of Sydney and the National ICT Australia Ltd., and gives seminars and a PhD course. The visit is funded by a travel grant award from the Swedish Foundation for International Cooperation in Research and Higher Education (STINT).
- *Jan. 2012:* Vangelis Angelakis starts his exchange activity in UK within a joint EC-FP7 funded Marie Curie IAPP project.

2011

- *Dec. 2011:* MT (Di Yuan) receives a grant award from the ELLIIT excellence center.
- *Nov. 2011:* Anthony Ephremides receives a grant award from the Swedish Research Council.
- *Oct. 2011:* MT (David Gundlegård, Di Yuan) receives a grant award from Vinnova within project Mobility of the Future.
- *Oct. 2011:* MT joins European Cooperation in Science and Technology(COST) action IC1101 "Optical Wireless Communications - An Emerging Technology", with Di Yuan being a national representative of Sweden in the Management Committee.
- *Sep. 2011:* Qing He joins MT.
- *Sep. 2011:* Lei Lei joins MT.
- *Aug. 2011:* Scott Fowler receives an EC-FP7 Marie Curie CIG grant award, with Di Yuan as the Scientist in Charge of the project.
- *Aug. 2011:* Lei Chen starts his exchange activity in UK within a joint EC-FP7 funded Marie Curie IAPP project.
- *Jun. 2011:* Lei Chen and Sara Modarres razavi receive travel grant awards from EC COST Action IC0902 for participating in the 1st summer school of IC0902.
- *Jun. 2011:* Joyce Wu of Ranplan Wireless Network Design Ltd, UK, starts her 3 month visit in MT, within a joint EU-FP7 funded Marie Curie IAPP project.

- *Jun. 2011:* Arash Matinrad leaves MT to launch his industrial carreer, joining Ericsson AB in Kista.
- *Jun. 2011:* Professor Gregorio Martinez Perez from the University of Murcia, Spain, visits MT and gives a lecture on his recent research on Current Security and Trust trends in Green Networks.
- *May 2011:* Prof. Sherali Zeadally, University of the District of Columbia, Washington DC, USA, visits the MT group. (Host: Scott Fowler). Prof. Zeadally is a recipient of the 2011 Fulbright Specialist Award for his exchange activity at Linköping University, Sweden.
- *May 2011:* Lei Chen receives a scholarship from the Ericsson Research Foundation as complementary support for his planned exchange in UK.
- *Apr. 2011:* Sara Modarres Razavi presents her Licentiate Thesis.
- *Mar. 2011:* MT has organized the 2nd Nordic System and Network Optimization for Wireless (SNOW) Workshop (with KTH, CWC-University of Oulu and the University of Maryland) in Sälen, Sweden, with great success. The workshop attracted over 45 participants from 10 countries.
- *Mar. 2011:* Di Yuan begins working part-time at Ericsson Research in Kista, Stockholm, for a duration of 1 year.
- *Feb. 2011:* Professor Anthony Ephremides, University of Maryland, USA, starts his guest professorship at MT.
- *Jan. 2011:* Vangelis Angelakis receives an ELLIIT grant award for research on fundamental capacity analysis in wireless communication systems, starting Q4 2011.

2010

- *Dec. 2010:* Lei Chen presents his Licentiate Thesis.
- *Dec. 2010:* Di Yuan receives a grant award from the Swedish Foundation for Strategic Research (SSF) for workin gpart time at Ericsson in 2011.
- *Nov. 2010:* Sara Modarres Rezavi receives a travel grant award from the Knut and Alice Wallenberg Foundation.
- *Nov. 2010:* Erik Bergfeldt defends his PhD thesis.
- *Oct. 2010:* Di Yuan gives lectures in Euro NF PhD course "Mixed Integer Programming in Survivable Telecommunication Network Design and Routing", Warsaw, Poland.
- *Oct. 2010:* Arash Matinrad joins MT.
- *Jul. 2010:* Johan M Karlsson becomes the coordinator of PhD study at the Communication and Transport Division.
- *May 2010:* David Gundlegård moves for 3 months to San Francisco, California, within an ongoing research collaboration with UC Berkeley.
- *Mar. 2010:* Scott Fowler receives a grant awardfrom ELLIIT for initiating research in QoS in LTE.

- *Mar. 2010:* Scott Fowler joins MT.
- *Jan. 2010:* MT becomes part of the ELLIIT excellence center funded by a Swedish governmental strategic grant.

2009

- *Oct. 2009:* Di Yuan offers the PhD course "Discrete Optimization with Application in Wireless Networks" at Politecnico di Milano, Italy, funded by the EuroNF NoE.
- *Oct. 2009:* Vangelis Angelakis joins MT.
- *Sep. 2009:* Di Yuan gives a plenary talk titled "Operational Research in the Design of Cellular Networks", at the 10th Symposium on Operational Research in Slovenia (SOR '09), Nova Gorica, Slovenia.
- *Jul. 2009:* Di Yuan moves for 3 months to the UK within an EC FP7 project.

GROWTH OF TELECOMMUNICATIONS

New and advance communications technology has resulted in an increased range of services available both to individuals and businesses. With over 46 million mobile phone users in the United Kingdom in 2002, modern telecommunications systems bring significant economic and social benefits. However the Council is conscious of the need to strike the proper balance between environmental objectives and sustainable telecommunications development.

EXISTING TELECOMMUNICATIONS NETWORK

Mobile telecommunications require the use of radio systems. Radio systems need aerials and towers to boost signals. The first mobile phone network was an analogue service which was phased out in 2001. Currently the predominant UK mobile telecommunication network is based on the 1990s second generation digital cellular Global System for Mobile Communications (GSM). This network is maintained by four licensed operators; O_2 (formerly BT Cellnet), Orange, T-Mobile and Vodaphone. Each provides a 98 per cent coverage of the UK. Two other service providers also provide a 2G service ('3' and 'Virgin') by using this network.

The continued growth in customer demand for second generation services (2G) i.e. mobile digital telephony, is placing stresses on the capacity of the networks. To accommodate growing customer requirements in terms of improvements to the quality of service delivery, provision of short message services (SMS aka Texting) and traffic handling capacity, many cells of the individual operator networks will need to be enhanced with additional base stations and masts.

Third Generation Network

Progress in telecommunications technology has resulted in the development of a new international standard and services. Known in the UK as the third generation mobile phone system (3G), in April 2000, five Universal Mobile Telecommunications Services (UMTS) licences were granted. The new licence holders include the four existing 2G operators and Hutchinson 3G. Improved data handling capacity means that 3G technology will enable the viewing of pictures and video and other enhanced services.

Under the terms of their licence each 3G operator is required to provide network coverage of 80 per cent of the population by 2007. Much of the existing infrastructure can be re-used although the new entrant, Hutchinson 3G, is required to build a network from scratch. The higher frequency ranges used by 3G technology means that the area coverage of individual cells will be reduced. This will result in an increased demand for new sites for masts and base stations. The Electronic Communications Code Operators have adopted a Ten Commitments framework which is enclosed in Appendix 3.

Other Networks

In addition to the new 3G licences, TETRA (Terrestrial Trunked Radio) has been developed for use as a secure network for such users as the emergency services. TETRA networks provide an integrated, digital mobile communications system. O_2 is currently developing a national network for the police called Airwave. Many new sites will be required for this network including some in environmentally sensitive areas.

PLANNING POLICY OF MOBILE COMMUNICATION

National Planning Policy

Planning Policy Guidance Note 8 (PPG8), published in April 2001, sets out the Government's planning guidance on the siting and design of telecommunication systems and installations. These cover a very wide range of systems including picocells (burglar alarm box size antennas), microcell and macrocell mobile phone base and relay stations and aerial masts.

PPG 8 offers guidance on environmental and health issues and consultation procedures. The guidance can be a material consideration in the determination of planning applications, and in the prior approval process and appeals.

The Office of the Deputy Prime Minister in November 2002 published a revised Code of Best Practice on Mobile Phone Network Development. This code was produced jointly by representatives of central and local government and the telecommunications industry. It provides advice on procedures, siting and design. It brings up to date best practice guidance and provides practical advice to facilitate improved communication and consultation between local operators, local authorities and local people. It aims to standardise practice thereby promoting greater consistency of approach and aid the transparency of the process for all concerned. This SPG by highlighting London Borough of Hounslow issues provides a local context for mobile telecommunication development.

The Mayor's Spatial Development Strategy for Greater London

The London Plan (February 2004) recognises the significance of modern communication technologies in enabling London to maintain its economic and cultural status as a World City. The Mayor supports measures to minimise the short-term social, economic, transport and environmental costs of introducing e-infrastructure including ducting and wireless equipment. There are already more Londoners in the teleworkforce than in any other European region. It is hoped that increased teleworking will reduce the need to travel thereby having a positive bearing on the fundamental issue of sustainability for London.

The Hounslow Unitary Development Plan

The London Borough of Hounslow Unitary Development Plan (December 2003) includes Policy ENV-B.1.7 which seeks to strike a balance between the environmental impacts of telecommunications development whilst recognising the Government's general policies of facilitating the growth of new and existing telecommunication networks.

Policy ENV-B.1.7 Telecommunications

Any proposal for telecommunications development will be assessed in terms of its operational requirements and impact on the local environment, and should satisfy the following criteria:

- The siting and design of the telecommunications equipment should cause the minimum visual impact on the local environment;
- Existing facilities should be shared wherever possible;
- Particular account should be taken of environmental policies relating to Listed Buildings, Conservation Areas, High Buildings, Green Belt, Metropolitan Open Land and other sensitive areas.

The Council is concerned about the visual impact of telecommunications equipment on the environment and aims to prevent a proliferation of structures, i.e. satellite dishes, microwave antennas, radio masts, aerials, etc, and a decline in environmental quality. Particular concerns relate to the siting and design of telecommunications equipment, sharing facilities wherever possible and taking account of other UDP environmental policies. Policy ENV-B.1.7 will be used to determine all prior approval applications.

OPERATIONAL NEED AND COMPETITION

PPG 8 states that local authorities should not 'seek to prevent competition between different operators and should not question the need for the telecommunications system (i.e. the individual network) which the proposed development is to support'. However PPG 8 considers it appropriate for planning authorities to request evidence from operators regarding the need for a specific proposal forming part of the overall network.

Prior Approval and Full Planning Applications should be accompanied by evidence of the need for the development in terms of the network coverage and/or network capacity. If an operator is seeking to increase capacity, a demonstration of existing traffic levels and evidence of the need for extra capacity (for example overloading circuits), would need to accompany the application as justification for the proposal. Similarly when seeking to increase network coverage, an operator would need to supply similar suitable evidence with the application.

PRE-APPLICATION DISCUSSIONS AND CONSULTATIONS

In accordance with the advice in PPG 8, the Council will undertake pre-rollout and pre-application discussions with operators. Similarly in line with PPG 8 and the Code of Best Practice, operators are encouraged to provide the Council with their annual rollout plans and to inform the Council of changes as they become available.

Operators will also be expected to consult the Council before the submission of prior notification or full planning applications in order to ensure that the optimum solution for each case is achieved.

At the pre-submission consultation stage, operators should provide basic information for proposal sites including site maps, outline plans, photographs and details of the type of installation.

Operators should demonstrate that the proposed installation is the minimum possible size and that the output is commensurate with effective service provision. At the time of notification or submission of an application, operators should be able to show the level of community consultation and discussions with other stakeholders which may have already taken place.

In accordance with the Code of Best Practice, telecommunications operators are encouraged to use the 'Traffic Light Model' developed by the Federation of Electronic Industries (FEI) and telecommunications operators. This model takes into account community, environmental and planning issues in relation to site selection.

Where a proposal is on or near a school or college, the operator, in accordance with PPG 8 should consult with the appropriate educational bodies. Paragraph 60 of the Code of Best Practice states the following factors should be taken into account when determining whether a school or college should be consulted:

- The proposed site is on school/college grounds
- The proposed development would be seen from the school/college grounds

- The site is on a main access point used by pupils/students to the school/college
- There is a history of concern about base stations within the local community
- The local planning authority has requested consultation with the school/college
- The school/college has requested that it be included in any consultation

The Council will expect the network operators to provide evidence of all consultations undertaken and will undertake any further publicity as may be considered necessary.

Supporting Information

The Code of Best Practice states that; 'The quality of information submitted as part of an application ...should always be clear and complete. Good quality submissions can help explain to local people and consultees as well as officers and elected members exactly what is being proposed and its likely impact ' and result in speedier decisions. The Code of Best Practice Annexes F, G and H, derived from PPG 8, identify the information which should accompany both prior approval and full planning applications in the London Borough of Hounslow.

LOCATION AND SITING

When assessing an application for telecommunications development, be it for prior approval or for planning permission, there will be two main areas of assessment: location and design and appearance. The main aim is to minimize the environmental intrusion of telecommunications development.

Environmental Considerations

The Council will have to determine whether the need to fill in the deficiency in a network outweighs any harm to the landscape or the environment. Protection from visual intrusion will be an important consideration in determining applications.

The sensitivity of the location of the proposal site for the telecommunications mast is a material consideration. Hounslow's flat topography may require more ground based aerial masts and antennae and positioning of antennae on high buildings than in areas with a more varied terrain.

Green Belt and Metropolitan Open Land

As an outer London suburban Borough, Hounslow's environment benefits from significant sections of Green Belt and Metropolitan Open Land (MOL). Planning Policy Guidance note 2 (PPG2) provides the national planning guidance with regard to Green Belt Land. PPG2 considers that inappropriate development is, by definition, harmful to the Green Belt. 'The visual amenities of the Green Belt should not be injured by proposals for development within or conspicuous from the Green Belt which, although they would not prejudice the purposes of including land in Green Belts, might be visually detrimental by reason of their siting, materials or design.' PPG8 (para. 65) further considers that telecommunications development in Green Belts is likely to be inappropriate unless it maintains openness. The applicant will need to demonstrate very special circumstances that would outweigh any harm to the Green Belt or MOL. (Green Belt Policies ENV-N.1.1 – 1.4 and MOL Policies ENV-N.1.5 -1.9). Such applications would normally be treated as a departure from the Plan.

Nature Conservation Areas

Hounslow has two Sites of Special Scientific Interest (SSSI). These are at Tide Meadow at Syon Lane and Kempton Nature Reserve. There are a further thirty nature conservation areas of Metropolitan and Borough Importance. Key UDP polices are ENV-N.2.1 (Strategic Nature Conservation Sites), ENV-N.2.2 (Sites for Local Nature Conservation) and Table ENV-N.2 (Nature Conservation Sites in the London Borough of Hounslow). Development adjoining strategic nature conservation areas will only be permitted where there would be no damage to the nature conservation interests. English Nature will be consulted on applications which would affect an SSSI. The Greater London Authority will be consulted on applications affecting nature conservation areas of Metropolitan Importance, and may be consulted on those affecting nature conservation areas of Borough Importance.

Historic Parks

In addition to Green Belt/MOL and Nature Conservation Sites, there are a number of Historic Parks and Gardens in the Borough. Such statutorily designated landscapes may also be Green Belt, Metropolitan Open Land and Nature Conservation sites. They provide landscapes of historic importance, identified by English Heritage and the setting of such protected areas should be respected and protected from visually intrusive development, including telecommunications development. This impact can be felt both singly and cumulatively.

These sites are listed in Policy ENV-N.1.16 (Historic Parks and Gardens) and include Chiswick House and Grounds, Hogarth House, Osterley Park, Strawberry House, Syon Park, Walpole House, Gunnersbury Park, Hanworth Park and Boston Manor. Furthermore it is considered important that development in this Borough should not have an adverse effect on the setting or views of historic parks and gardens in adjacent boroughs. When considering a telecommunication proposal which would affect an Historic Park or Garden the applicant should consult the Council's Conservation and Design Officer (020-8583-4941). English Heritage and the Garden History Society will be consulted on any such application.

Tall Buildings

In the case of proposals to locate masts and antenna on tall buildings, in addition to other policy considerations, regard will be given to UDP Policies ENV-B.1.2 (High Buildings and structures affecting sensitive areas) and ENV-B.1.3 (High Buildings or structures in areas other than those listed in ENV-B.1.2).

Listed Buildings

There are more than 800 Listed Buildings in Hounslow, details of which are kept by the Council on a database. UDP Policies ENV-B.2.5 (Development Affecting the Setting of a Listed Building), ENV-B.2.6 (Identification and Protection of Buildings of Local Townscape Character) and ENV-B.2.7 (Alterations to Listed Buildings and Buildings of Local Townscape Character) aim to ensure that the setting and character of these buildings and structures is not harmed, that any scheme is both well designed and sympathetic and that the works are justified. Any application involving a Listed Building will require Listed Building Consent in addition to a Full Planning Application. It is recommended that an applicant should consult the Borough's Conservation and Design Officer (020-8583-4941) when considering a proposal which would affect a Listed Building or building of Local Townscape Character. English Heritage will also be consulted where appropriate.

Conservation Areas

There are more than twenty conservation areas in Hounslow. Policy ENV-B.2.2 (Conservation Areas) details the policy framework for conservation areas with further information in Supplementary Planning Guidance. The unique character of these areas means that special attention will be paid to their preservation and enhancement and in particular any harm caused by new development. It is recommended that when considering a telecommunications development in a conservation area, the applicant should contact the Conservation and Design Officer (020-8583-4941).

Heathrow Airport

With Heathrow airport to the west of Hounslow, much of the Borough is within a 3km radius of the airport. Consequently, the applicant will need to notify British Airports Authority (BAA) and the National Air Traffic Services Ltd (NATS) for the installation of a mast within this 3km area, as a requirement of their network operator licence. Moreover, there are some areas beyond this range which would require BAA and NATS to be consulted for development proposals with heights above ground level exceeding 45m.

Residential Areas

When considering the location of masts in or adjoining residential areas, operators should demonstrate that nearby industrial or other areas are unsuitable and that the need for the development outweighs the harm to the visual amenities.

In addition to the designated area/special issues key site considerations also include:

- Height of site in relation to surrounding area
- Existence of topographical features/natural vegetation/ archaeological features
- Site prominence from any side/vista
- Site in relation to existing mast/ structures/ buildings. Need to avoid clutter
- Landscaping and Screening

When a building or structure supporting telecommunication equipment is demolished/replaced, any new/replacement installation will need to be considered by the Council either as a prior approval determination or as a planning application.

8

Careering Through the Web and Mobile Communication

APPLICATIONS OF WEB TECHNOLOGIES

Given the wide range of technological developments that have the potential to impact on career exploration processes, it is important to examine how these have been being incorporated into the practices of a range of organisations. This paper demonstrates that technology has generally been used to help meet client demand in one of three ways:

1. To deliver information;
2. To provide an automated interaction; or
3. To provide a channel for communication. Where technology is used to deliver information, it serves a range of functions.

It can, for instance, recreate the careers library by supplying information about jobs and courses. This can increase clients' access and remove the space limitations that plagued the conventional careers library. However, this kind of technology also provides an opportunity to improve the quality of information, to harness the linked nature of the web to draw in external resources (such as employers' sites) and to provide a more media-rich experience through the use of pictures, audio and video. Where technology is used to develop an automated interaction, there are a range of opportunities. The use of technology can automate the initial exploration and diagnostic elements of the usual advice and guidance service: for example, it can facilitate psychometric, matching and reflective tools, and perform some initial diagnostic tests.

Technology can also be used to support people to develop their career learning skills: for instance, through games and simulations that can be used to provide an interactive way of exploring the worlds of learning and work. These technologies therefore both promote user-control and self-reliance, but also can automate some of the more routine aspects of the guidance process, so allowing professionals to focus on offering higher level support to clients. Finally, there is a range of tools that facilitate communication and interaction between people, usually but not always at a distance. These technologies include the telephone and email as well as a range of technologies that enable telephone and email communications to be more effectively managed. Such technologies can make professional support to individuals more accessible, and are also being used to build communities of learning. These technologies can be further classified as those that facilitate the following types of communication:

- One-to-one;
- One-to-many and many-to-one; and
- Many-to-many.

Each of these different forms of communication offers new potential for both career exploration and the delivery of career IAG.

POLICY ISSUES OF ACTIVITY AND OPPORTUNITIES IDENTIFIED

Given the range of activity and opportunities identified, it is difficult to make straightforward policy recommendations. This is particularly the case as the internet is notoriously difficult to regulate. However, if the online career exploration environment is conceived as a market, it is possible to draw on previous work relating to the development of a market in careers to suggest three roles for public policy:

- Stimulating the market in order to build its capacity.
- Regulating the market and assuring the quality of services, both to protect the public interest and to build consumer confidence.
- Compensating for market failure where this is appropriate. A fourth role might be added: supporting the growth of an educated consumer. In order to achieve this and to maximise the value that individuals can draw from the online market, it may be useful to consider policy recommendations in three areas:
  - Supporting the growth of digital literacy to support career management;
  - Developing mechanisms for quality assurance and recommendation of career-related online resources;
  - Supporting the technical upskilling of careers professionals.

Digital literacy clearly has a key role to play within an individual's pursuit of work and career. The ability to identify opportunities, gather labour market information and harness social networks are key to individual career success. Within this context, it is possible to conceive digital literacy as part of a wider set of career management skills. Careers services have a role to support and develop the digital literacy of their clients. Developing digital literacy in the context of career development is essential in equipping people to interact with the worlds of learning and work and therefore needs to be a stated outcome for careers services. While the ideal would be to have a highly digitally literate population, this is challenging to achieve. There is therefore a clear role for evaluation of resources and the provision of guidance around their use. It is possible to consider their role in a number of different ways. If all careers professionals were utilising social media to publish their insights and publicly identify the resources that they found useful, there would be an enormous critical mass of new content and metadata that would balance existing content (much of which is produced by the private sector and is highly US-centric). Finally, it is important to acknowledge that the picture of technological change that is painted throughout this paper is a complex one. There are technical, social and cultural elements to understanding the ways that new technology can support career exploration. Careers professionals do not have sole responsibility for all career learning or for the development of all online career exploration. They do however have an important role in supporting the development of career-related digital literacy, in quality-assuring career learning materials and in developing a strong understanding about the inter-relationship between technological development and the pursuit of career.

ROLES OF INTEGRATED WEBSITES

Offer & Sampson (1999) suggest that career centres' websites can have at least five different purposes. One is as a *funnel* into their own existing off-line services, aiming to maximise take-up of these services. The second is to act as a *diversion,* seeking to take the pressure away from these off-line services by diverting users to other, usually web-based resources where their needs can be met. The third is an enhancement of such diversion, seeking to deliver *on-line guidance* within the site itself. The final two add further enhancements: providing a *forum* for putting users in contact with others facing similar issues to their own, or with people who may offer help in relation to these issues (e.g. potential career mentors); and providing a source of *distance learning* programmes in career management skills and related areas (for examples, see Offer *et al.,* 2001).

Most of these purposes are not, of course, mutually exclusive, but the choice and balance between them require strategic decisions to be made, as do the choice of partners for any partnership arrangements. Not only this, but the process of making such decisions can be a valuable opportunity to review the strategic development of the centre as a whole. Many career centres clearly start by simply establishing a presence on the Internet: as Offer (1998) puts it, "we're here, because we're here, because we're here". Thereafter, it seems likely that they go through four developmental stages. The first is *promotional*: promoting what the service offers off-line. The second is *adaptive*: delivering some of these services in on-line form. The third is *innovative*: delivering new services on-line which are not possible, or less feasible, off-line – which is where any partnership arrangements may come into play. The fourth is *synergistic:* to intertwine on-line and off-line services in new ways. The further a career centre moves through these stages, the more it is using its website as an agent of change in relation to its service provision as a whole (Offer *et al.*, 2001).

TELEPHONE HELPLINES

Alongside websites, there has been growing interest in the use of telephone helplines in delivering career information and guidance services. The largest telephone helpline service in this field is the Learndirect service in the UK, which was launched in February 1998 and by the end of 2000 had responded to 2.4 million calls. Other more limited helplines have been launched in other countries, including Canada and New Zealand. A comparative analysis of these helplines indicates that some have been promoted essentially as *information* services; others as career *counselling* services. Some are focused primarily on *learning* or on *work;* others on *career*, embracing the two. Some are aimed at *young people* or *adults*; others are *all-age.* Some are *separate* services based on callcentres; others are *integrated* in various ways into more broadly-based services (Watts & Dent, 2002).

The decisions made about the framing of helplines in these various respects are critical. Included in this is the extent to which the service should be offered at national or at local level. With Learndirect, the original idea was that calls should be routed to helplines based as locally as possible. This is the model used, for example, by the UK National Health Service helpline NHS Direct: calls are routed to local callcentres, and are only passed elsewhere when lines are busy; the notion being that, in time, it might become the "gateway" to all local health services (McLennan, 1999). In the case of Learndirect, by contrast, the helpline is offered largely at a national level. This can result in some loss of quality, in terms of access to local knowledge. Against this needs to be set the consistency of service and of helpline adviser training offered by relatively large-scale operation. Such decisions have a significant influence on the balance between "global" and "local" dimensions in guidance provision. This in turn is critical in determining the level of integration that is possible with face-to-face services.

Technological synergy

Whereas the Learndirect helpline was originally conceived as a separate service, a Learndirect website has now also been introduced, and increasing attention is being focused on integration between the two. The website includes not only courses and occupations databases, but also a diagnostic package which provides an assessment of skills, interests and values, and connects the results to occupational families. Each page of the website includes a "call me" button which generates a telephone call from a Learndirect adviser. In principle, it should be possible in future for the adviser to bring up on their screen the caller's work to date – a draft curriculum vitae, for example – and work on it on-line with them. Conversely, more callers could be encouraged to access the website and be supported in doing so. Again, greater use of e-mail should make it possible to sustain contact over a period of time through a mixture of synchronous (e.g. telephone) and asynchronous (e.g. e-mail) communications.

The concept of flexible usage of the telephone, website and e-mail, linked with face-to-face facilities, opens up new opportunities for the delivery of career information and guidance. It means that individuals can initially access help in the form which is convenient to them and with which they feel comfortable. Some feel comfortable visiting a careers centre; some do not. Some are more comfortable on the telephone, or on e-mail; some are not. A further dimension will be added to this by the likely move towards ready domestic access to videophones or interactive digital television. All of these could be regarded not as alternative services but as portals into a wide, flexible and well-harmonised network of services which can enrich the learning pathways available to the individual. Public policy sometimes seems to impede such harmonisation; it could, however, proactively promote it.

Resource-centred v. relationship-centred

The significance attached to the relationship between the individual and the counsellor. The model could be based on a co-ordinated range of *resources*, of which the counsellor is seen as one. Or it could place the *relationship* with the counsellor at the centre, viewing other resources as supports to this relationship. In the latter case, it needs to be recognised that the relationship can now be sustained in a variety of ways: not only face-to-face but also at a distance; and in the latter terms, both through synchronised communications like the videophone, the telephone and Internet "chat", and also through asynchronous communications like e-mail and voice-mail.

The rationale for the relationship model is that career decisions have an important cognitive component, but that they are bound up very closely with people's feelings about themselves, their sense of identity, and their dreams and aspirations. Accordingly, the individual can best be helped by working in a relationship with another individual who has the skills to enable them to address their distinctive identity. This is a strong tradition within the career guidance field, but has come under attack on two linked grounds: that by placing the relationship at the centre, it cedes too much power to the counsellor; and that because it is so labour-intensive and counsellors are costly, it is not practicable or sustainable as an extensively delivered model. This leaves the alternative view that guidance professionals should see themselves primarily not as counsellors but as managers of guidance resources: managing diverse resources in ways which enable individuals to find the means through which their personal needs can best be met (Watts, 1996a).

In relation to these two models, ICT tends to be viewed as leaning towards the resource-based model. But this is not necessarily the case: certainly ICT does not of itself demand a jettisoning of the relationship model. Indeed, there is a risk that contrasting the two models too starkly may implicitly hold frozen the familiar and established practice of the face-to-face one-to-one interview. As Tait (1999) has pointed out, it is important to understand how technology is now transforming the ways in which human relationships are pursued and managed. Instead of assuming that crucial elements are lost when relationships are mediated by technologies, more attention needs to be paid to how technology is enabling such relationships to be developed and sustained in ways which are released from the constraints of space, time and physical presence.

Constantly lurking beneath professional anxieties about the use of ICT in guidance has been the fear of reductionism: that the use of technology will lead to simplistic, "quick-fix", information-based approaches, in which the human element is marginalised or eliminated. Increasingly, however, it can be used within an integrated approach not only to supplement but also to extend the range of this human element.

The key distinctions now are not simply between human interventions and ICT-based interventions, but between a more complex range of interventions: direct face-to-face counsellor-client interactions, on either a one-to-one or group basis; technically mediated counsellor-client interactions, on either a synchronous or asynchronous basis; and stand-alone ICT-based services in which standardised services can be repeatedly used by different users without additional human resource cost. An important issue is how these different interventions can be optimally blended in effective service delivery, attending to clients' varied needs and preferred learning styles. This has implications both for the training of counsellors, and for wider public policy.

TECHNOLOGY AND CAREER INFORMATION, RECOMMENDATION AND GUIDANCE

The sector should create a range of IAG services which utilise 21st century technologies is timely. The technologies often referred to as Web 2.0 (social networking, user-generated content, segmentable and repurposable data etc) offer substantial possibilities for changing the way people work, communicate and learn. Much thinking has already been done on the ways in which learning contexts are changed by Web 2.0. There has also been considerable discussion on the impact of these technologies in a variety of fields such as business (Tredinnick, 2006), social organisation (Shirky, 2008), medicine (Giustini, 2006) and the democratisation of information (Weinberger, 2008).

Career IAG professionals and researchers have long paid considerable attention to the implications of technological change for service delivery. There is a substantial body of research looking at the relationship between ICT and guidance. This literature has also examined key issues such as the role of the internet in IAG, e-guidance and developing the IAG workforce to make use of ICT. The rapid expansion of the internet and Web 2.0 applications has meant that this literature can date quite quickly. However, many of the overriding principles and issues from earlier study are still relevant to current practice. Watts (2002) identifies four phases in the development of ICT in guidance: the mainframe phase (1960 to late 1970s), the microcomputer phase (1980s to mid-1990s), the web phase (1990s to early 2000s) and the digital phase (current). He also identifies three trends that can be seen across the four phases: increased accessibility; increased interactivity; and more diffused origination (the trend towards more diverse creators and providers of career IAG). He notes that the role of ICT in guidance can be seen as a tool, as an alternative, or as an agent of change. While policy-makers have sometimes tended to view it in the first two guises, the main policy challenge is now its potential as an agent of change.

It is important that these web developments are seen in the context of wider technological changes that have the potential to impact on guidance. The development of telephone-based guidance services has been increasingly assimilated into the mainstream practice of guidance professionals and has been seen to deliver high levels both of customer satisfaction and of expert-rated quality. It is noteworthy that in England the number of telephone interventions has declined somewhat in relation to web-based interventions, but that the telephone is increasingly being used for in-depth guidance rather than information and advice. These trends are best seen in the light of the diversification of communication modes used by careers services and the increasing integration between web and telephone technologies. Thus the Careers Advice Service has now augmented its telephone guidance service to include text messaging and email, discussion forums, and blog-based services.

Another important strand of technologies that are used in guidance is ICT-based career guidance systems such as the various products produced by Cascaid and the Kuder Career Planning System (www.kuder.com). These systems typically use a mixture of interest and skills assessments to provide targeted occupational information and advice. The Association of Computer-based Systems for Career Information (ACSCI) exists to represent these systems and has developed some useful standards for their development. There is an extensive literature that examines the appropriate use and effectiveness of these systems. Many of them evolved iteratively through Watts's four phases (mainframe, microcomputer, web, digital). How these systems are interacting with the Web 2.0 world. Cogoi (2005) explored the range of technologies that were being used to support guidance work and concluded that there was a distinction between two types of uses:

- As a *medium* for the interaction between a guidance worker and a client. This group of technologies included the telephone, email, chatrooms, newsgroups, SMS (text messaging) and video conferencing.

- As resources to provide information and support client decision making.

This remains a useful distinction. However, there is a need for a more complex typology to take account both of the enhanced opportunities offered by new technologies and of the developing practice of careers practitioners.

In this paper the focus is to explore the potential of new technologies, and the potential that new developments create for enhanced access to information, to communication and to social interaction. Its purpose is to explore the ways in which these might be adapted by careers services or used by people to support their career learning, decision-making and progression. It uses examples from current practice to illustrate particular points or to identify areas where technologies exist but their application in a careers context does not. It is not a comprehensive audit of technological innovations within the guidance sector; neither is it a critique of current uses or services. Rather it uses current practice to explore the potential for future career service development in an increasingly digital world.

THE POTENTIAL OF NEW TECHNOLOGIES TO MAINTAIN CAREER DEVELOPMENT

Key to conceptualising the impact of new technologies on individuals' career development processes is a consideration of the demographics of web use. In 2009, 70% of UK households had internet access, and 76% of adults had accessed the internet in the previous three months. Amongst young people (16-24) surveyed in 2008, 93% had accessed the internet in the previous three months, and 77% accessed the internet every day or almost every day (Office for National Statistics, 2009b). All of these figures are trending upwards, as are figures for use of social media, mobile phones and figures relating to the use of the internet by children.

While the amount and the ways that people use the internet clearly have a relationship to age, this relationship is not always as straightforward as is often perceived. Ofcom (2008) found that the use of social networking sites was well distributed across the age ranges. Thus, for example, 49% of members of social networking sites were over 35, and 24% were over 50.

As the internet becomes increasingly ubiquitous, it is easy to ignore the dwindling group who are not connected or engaged. Digital exclusion is not simply about the inability to find an internet connection, but rather relates to broader issues of culture and skill. It is argued that there is a particular combination of skills, attitudes and knowledge that can be characterised as digital literacy. There is an exponential and positive relationship between an individual's digital literacy and the value that they can extract from the internet. But it is clear that while internet use might be increasingly widespread, digital literacy is not equally distributed throughout society. Those who are at greater risk of digital exclusion may include those from working-class backgrounds, the elderly (Helsper, 2009) and the socially excluded – for example, the rural poor (Warren, 2007).

However, digital exclusion cannot simply be assumed to be a function of someone's age, geographical location or place in society, as Roberson & Nardi's (2010) work on the use of digital technologies by homeless people in Los Angeles shows. Roberson & Nardi demonstrate that some homeless people are digitally literate and able to gain access to the web via devices provided by support services or using their own mobile phones or laptops; where the homeless were digitally included, it enabled them to use these skills to support their survival needs and to increase their social capital and social inclusion.

It is possible to take this discussion beyond simply identifying those who use the web and those who do so less or not at all. The question of who uses the web is necessarily followed by a discussion of how they use the web. The BBC is currently conducting an experiment with Ian Rowlands of University College London into online behaviour. Rowlands (2010) argues that "we know alarmingly little about how people really interact with the web" and poses a series of questions that are likely to differentiate different types of web user:

- How able are we to filter the enormous amount of information delivered to us by search engines?
- How confident are we in our ability to find the "right answer" when looking for information on the web?
- How much do we concentrate when we are using the web, even when we are using it to make potentially life-changing decisions?
- What is the role of memory in shaping what we do on the web? (How much of today's online journey do you actually remember?)
- How influential are "super brands" in defining the choices we make online?
- How do we decide on the reliability or "authority" of information provided by others online?

While it is unlikely to be possible for the providers of careers information and guidance to be able to optimise their sites for a wide variety of web learning styles, it is worth noting that there are likely to be significant differences in how individuals interact with web-based content. These may have a complex series of relationships to factors like age, educational level, digital literacy and personality.

The complexity of issues around the digital divide can make it difficult to formulate policy in this area. It would be easier if we could comfortably say that everyone had access to the internet or that certain groups were fully excluded. While there is some truth in the idea that those experiencing social exclusion are less likely to be digitally literate and included, this is by no means a rule. So far, careers services working with high-skill, socially-included individuals, for example in higher education, have been most able to grasp the potential of new technologies in careers work. However, some members of socially excluded groups such as NEETs (Marschalek & Unterfrauner, 2009) and the long-term unemployed are internet or web users. Marschalek & Unterfrauner's work on socially excluded young people suggests that many of the benefits of internet access are available to this group through mobile technologies.

Furthermore, there is a growing body of work looking at the potential of the internet to re-engage the socially excluded in the labour market. Kuhn & Skuterud (2004) found that while unemployed workers who used the internet for job search were more likely to be successful than those who did not, this was no longer true when they controlled for other variables. McQuaid, Lindsay & Greig (2003) concluded that while use of ICT for job-seeking was growing amongst unemployed workers, it was still at a relatively low level; they also found a tendency for the rural unemployed to be more engaged with internet and telephone technologies for job-searching. However, both of these studies are now more than five years old, and it is possible that Marschalek & Unterfraunder's findings might now be more widely generalisable, given the increasing level of convergence between web and telephone technologies.

As can be seen, technology is developing rapidly, so it can be difficult to evaluate comprehensively what might be useful for career development work. This section will now go on to outline a series of technological trends that seem to raise important opportunities or issues for careers work. These trends are based on the authors' experience, discussion with career practitioners and a review of literature relating to recent technological trends.

IMPACT OF THE INTERNET

In *Bowling Alone*, Putnam (2000) argues that the key issue in assessing the impact of the internet is whether people use it more like the television (alienating, individualised, based around consumption) or more like the telephone (connecting, social, producing). It now seems clear that the internet has become a super-charged telephone at least as much as it has become a multi-channelled television.

A large number of the opportunities associated with new technologies are about finding new ways for people to talk to each other, whether one-to-one (chatrooms, videoconferencing etc), one-to-many (blogs and video-sharing sites) or in groups and communities (social networking and community communication sites). This myriad of new ways in which people can communicate is driving social and political change (Shirky, 2008) but is also increasingly challenging the way in which people use the web and discover information.

To a growing extent, individuals are discovering internet resources via peer recommendation. It has been observed that this is starting to challenge search engines as the main way in which resources are discovered (Long, 2010). Search engines can only provide the answer to a question that you already have, while social networks have the capacity to provide new and surprising questions. This ability to provide challenging and new information offers huge potential for career exploration, as it has the potential to provide options and ideas for exploration that may not previously have occurred to the individual.

In exploring the value of technology for career IAG, it is therefore important to engage with these increased opportunities for human interaction. Careers websites have tended to see the web largely as a broadcast (one-to-many) medium through which information can be communicated from the professional to the client. Where communicative technologies were used, they were similarly conceived as an extension of the existing one-to-one practice of career IAG (Offer, 2004). The technologies associated with the social web challenge this by increasingly utilising many-to-many forms of communication. This requires a new kind of guidance pedagogy that recognises and utilises the socially situatedness of communications, perhaps by seeking ways of engaging peers, mentors and employers more strongly in the guidance and career decision-making process.

Collectivising knowledge

Current technologies allow us to harness collective intelligence in ways that radically alter the way we understand the role of expertise and the production of information. O'Reilly (2005) contrasts the approach which underlies Amazon's search with that which underlies its competitor Barnes and Noble (barnesandnoble.com). Amazon's listings always favour the "most popular", aggregating the choices of its customers, whereas (at the time of O'Reilly's article) Barnes and Noble used their website to push their own products and priorities. However, the idea of harnessing and collectivising knowledge goes beyond this kind of plebiscite democracy by creating ever more sophisticated mechanisms for aggregating the opinions of the many. A system like Wikipedia enables its users to engage in the creation of a negotiated understanding of truth, one which has been found to deliver more reliable content than its rival the expert-produced *Encyclopaedia Britannica* (Giles, 2005). This is not to say that there is no value in expertise. But, as Surowiecki (2005) argues in The Wisdom of Crowds, if you can find a way to aggregate the opinions and knowledge of the crowd, this can often outperform conventional expertise.

If Wikipedia socialises the creation of data, services like Flikr (www.flickr.com) and Delicious (delicious.com) socialise the creation of metadata. In a world where there are no real limits on who can publish or the amount they can publish, the organisation and categorisation of information becomes even more important than the information itself. Conventionally, this was done through the creation of an expert taxonomy: an example in the careers field was the Connexions Resource Centre Index (Connexions, 2008). Web 2.0 technologies enable these conventional expert-designed taxonomies to be replaced with folksonomies. A folksonomy is an organisational structure defined by users rather than by experts or designers. It allows multiple personal structures to be created, as well as offering the capacity to aggregate these subjective individual structures together into a collective metadata structure. So, for example, a search for "career" on the social bookmarking site Delicious lists 310,246 sites relating to career (2nd April 2010). Users can use this list to find out the sites that have been tagged as relating to career most often by others, and to access a series of related terms. So we can see that over 17,000 people see the business networking site LinkedIn as having relevance to career, and that 2,306 people have tagged a post on Bhuvana Sundaramoorthy's Blog entitled 50 Common Interview Q&A. It also demonstrates that the terms "job", "jobs", "business" and "resumé" are commonly seen as related to the term "career."

These new ways of aggregating knowledge support the development of a public sphere within which ideas can be shared, debated and synthesised, and reinforce the trend towards community. Surowiecki (2005) has theorised the challenge that markets and other aggregators of information can pose to conventional ideas of expertise. He argues that, given the right conditions and a functional system for aggregation, a crowd is likely to make better decisions and to possess more knowledge than an expert. Surowiecki's conditions are that the crowd needs to be diverse, that its members reach decisions independently, and that they work on the problem in a decentralised way.

While is possible to dispute Surowiecki's thesis (Tammet, 2009), it is difficult to challenge the idea that the internet offers a wide range of tools to aggregate the views of the crowd and that these aggregations are becoming increasingly influential. The generation of huge amounts of user content and its mediation through equally large amounts of user metadata mean that services which are based heavily on the possession of information, or on experts' ability to organise and locate this information, may need urgently to rethink their unique selling points.

The ability to use technology to aggregate the knowledge of individuals offers a range of exciting possibilities for career exploration. Wikijob (www.wikijob.co.uk) provides one example of this. It offers stakeholders a career learning resource which is developed and written by users of the site, on the same basis as Wikipedia. The fact that contributors include career learners, career professionals and employers has the potential to strengthen the site.

The successful generation of a community of editors expands its scope to develop and respond to change beyond what would be possible within a single organisation. However, the challenge to develop a sufficiently large community of authors and editors (as has been done with Wikipedia and possibly with Wikijob) remains the key challenge.

Individualisation

The ability for a user to control their own interface with web content is another important theme. Previously, organisations have seen their websites as the centre of their online communications with their users. Effort was expended to make sites "sticky", so that once you entered you would not want to leave. However, the increasing powers of search mean that users have much more capacity to dive straight into the content that they want. This has the potential to re-conceive how interactions happen on the web. Whereas energy has hitherto been expended on the creation and organisation of the interaction space (the website), it is increasingly the case that the interaction space is the internet as a whole. In this case it may be that the traditional website becomes less important as users engage with granulated content through a range of platforms.

Aggregation is an important user strategy to manage interactions with a wide variety of web content. It includes a variety of services that allow users to bring together material which interests them, to help them to assimilate it, and to build relationships between different entities. The development of portals (applications that co-locate data or functionality from two or more external services), aggregators (applications that create a single datastream from multiple sources) and mashups (applications that combine data or functionality from two or more external sources to create a new service) all enable the user to individualise their interactions with individuals, organisations and information.

Individualisation provides a powerful way for individuals to manage information overload, but its potential goes beyond merely filtering out information. The ability to create a personalised interaction with online information paves the way for the creation of what has been described as a personal learning environment (PLE) (Attwell, 2007). The PLE is not a piece of technology but rather a type of interaction that can utilise a range of technologies. Johnson & Liber (2008) argue that the development of user-driven PLEs necessitates a new pedagogy which recognises this changed dynamic between individuals, professionals and institutions.

For careers organisations, the main issue is to provide content in ways that support the creation of individual's learning and information pathways through their content. The use of push technologies like RSS which are designed to syndicate content to users are likely to be an important component of this. Careers adviser Elizabeth Wilkinson has used Netvibes to draw together a range of resources for her clients (postgraduates at the University of Manchester) on her Resources for Postgraduate Careers page (www.netvibes.com/postgradcareers#The_ Careers_Service). This serves the dual purpose of providing an aggregated point of access to a range of services, and amplifying an approach to information management that clients can use. In order for individuals to access fully the opportunities offered by individualisation, they will need to increase their ability to manage and interrogate information.

situated in the cloud

The cloud is a metaphor that is used to describe the intangible and distant nature of the internet. Whereas information that is stored on your computer sits in a black box next to your screen, information that is stored on the internet has a cloud-like existence, at once observable and present but also intangible and distant. It represents a paradigm shift in computing, from a situation where most of your data and applications were located on your computer, to a situation where most data and applications are delivered via the internet. So we see a move from desk-based applications like the Microsoft Office suite to cloud-based applications like Google docs, which provides a range of office-based applications and the capacity to store your data online. For programmers and developers, cloud computing goes beyond providing a way to store data or access applications, and opens up a new scalable environment within which development can take place. For organisations, cloud computing has big implications for the management of technical infrastructure. The locus of infrastructure moves away from a single organisation towards the internet. However, cloud services do not really live in the cloud, but are usually supported by a third-party supplier with a huge data centre who can provide infrastructure, applications and/ or space in a similar way to the way a utility is provided. From a business perspective, the key advantage therefore lies in the ability to scale a range of technical resources up and down to fit with demand.

For individuals, cloud computing opens up some possibilities that are likely to have implications for career. The ability to create online spaces which can serve as repositories for data, to move data easily between learning and work environments, and to be able to share these resources with others, has the potential to mainstream the idea of the e-portfolio. This builds on the discussion of PLEs in section 3.3 by enabling individuals not only to engage with a range of personalised learning content but also to respond to it and share it with both professionals and their peers. E-portfolio tools are now being used strategically by a range of careers services to facilitate career learning (Watts, 2010), so it is interesting to investigate ways in which use of reflective career learning processes can be extended beyond the formal educational contexts within which e-portfolios are usually introduced. The ability to create, organise and share materials can support and smooth transitions and can provide ways to record and market individual and collective achievement. E-portfolios within careers-service websites provide one way to achieve these benefits, but it is increasingly possible to "portfolio-ise" your learning and experience using tools such as LinkedIn (to provide CV-type material, qualifications and references), blogs (to provide textual examples of your work and thinking) and Flickr or YouTube (to provide visual and multi-media content). The cloud simply describes the location of these materials; the ability to use this location to selectively retain and broadcast key life, learning and work information across the course of a life journey has huge potential implications for career development.

CAREERING THROUGH THE WEB: PROFESSIONALS AND RESEARCHERS

Career IAG professionals and researchers have long paid considerable attention to the implications of technological change for service delivery. There is a substantial body of research looking at the relationship between ICT and guidance. This literature has also examined key issues such as the role of the internet in IAG, e-guidance (Madahar & Offer, 2004) and developing the IAG workforce to make use of ICT.

The rapid expansion of the internet and Web 2.0 applications has meant that this literature can date quite quickly. However, many of the overriding principles and issues from earlier study are still relevant to current practice. Watts (2002) identifies four phases in the development of ICT in guidance: the mainframe phase (1960 to late 1970s), the microcomputer phase (1980s to mid-1990s), the web phase (1990s to early 2000s) and the digital phase (current). He also identifies three trends that can be seen across the four phases: increased accessibility; increased interactivity; and more diffused origination (the trend towards more diverse creators and providers of career IAG).

Key to conceptualising the impact of new technologies on individuals' career development processes is a consideration of the demographics of web use. In 2009, 70% of UK households had internet access, and 76% of adults had accessed the internet in the previous three months (Office for National Statistics, 2009a). Amongst young people (16-24) surveyed in 2008, 93% had accessed the internet in the previous three months, and 77% accessed the internet every day or almost every day (Office for National Statistics, 2009b). All of these figures are trending upwards, as are figures for use of social media, mobile phones and figures relating to the use of the internet by children (Office for National Statistics, 2009c).

As the internet becomes increasingly ubiquitous, it is easy to ignore the dwindling group who are not connected or engaged. Digital exclusion is not simply about the inability to find an internet connection, but rather relates to broader issues of culture and skill. It is argued that there is a particular combination of skills, attitudes and knowledge that can be characterised as digital literacy. There is an exponential and positive relationship between an individual's digital literacy and the value that they can extract from the internet. But it is clear that while internet use might be increasingly widespread, digital literacy is not equally distributed throughout society. Those who are at greater risk of digital exclusion may include those from working-class backgrounds (Tien & Fu, 2008), the elderly (Helsper, 2009) and the socially excluded – for example, the rural poor (Warren, 2007).

Access to online technologies has continued to grow. While it is still not possible to talk about ubiquitous access, the convergence between web and telephone technologies means that a very high percentage of the UK population can now access online services if they want. This is not to say that the digital divide has vanished altogether, but rather that the divide is now principally centred around issues of culture, education and skill rather than the capacity to access an internet-capable device. It is more helpful to conceive this as the development of digital literacies: skill-sets that are needed to effectively utilise online resources. The poorer, more socially-excluded and older members of society are likely to have lower digital literacy, though it is important not to over-generalise. Given the importance of the online environment to career exploration, it is vital that the development of digital literacy be given a high priority.

TECHNOLOGICAL TRENDS

We have explored a range of technological trends that are likely to have impact on the way this career exploration environment develops. The eight trends we have identified are:

- *Community:* The internet is an important site for community interaction: technology has increasingly become a tool to facilitate a wide range of communication. This myriad of new ways in which people can communicate is driving social and political change (Shirky, 2008) but is also increasingly challenging the way in which people use the web and discover information. To a growing extent, individuals are discovering internet resources via peer recommendation. The technologies associated with the social web challenge this by increasingly utilising many-to-many forms of communication. This requires a new kind of guidance pedagogy that recognises and utilises the socially situatedness of communications, perhaps by seeking ways of engaging peers, mentors and employers more strongly in the guidance and career decision-making process.
- *Collectivising knowledge:* A feature of many Web 2.0 technologies is their ability to collectivise and aggregate the opinions of many. Current technologies allow us to harness collective intelligence in ways that radically alter the way we understand the role of expertise and the production of information. Web 2.0 technologies enable the conventional expert designed taxonomies to be replaced with folksonomies. A folksonomy is an organisational structure defined by users rather than by experts or designers. It allows multiple personal structures to be created, as well as offering the capacity to aggregate these subjective individual structures together into a

collective metadata structure. These new ways of aggregating knowledge support the development of a public sphere within which ideas can be shared, debated and synthesised, and reinforce the trend towards community.

- *Individualisation:* Users are increasingly able to individualise and tailor their relationships with online content. Individualisation provides a powerful way for individuals to manage information overload, but its potential goes beyond merely filtering out information. The ability to create a personalized interaction with online information paves the way for the creation of what has been described as a personal learning environment (PLE) (Attwell, 2007). The PLE is not a piece of technology but rather a type of interaction that can utilise a range of technologies. Johnson & Liber (2008) argue that the development of user-driven PLEs necessitates a new pedagogy that recognises this changed dynamic between individuals, professionals and institutions.
- *Recognising time and place:* Technologies are now enabling us to interact with the web in ways that recognise and identify time and place. This is particularly important to career exploration. Asking a computer to identify what is recent is difficult, but asking it to identify what is current is more difficult still. Similarly, identifying where something originated, where it relates to, and how broad a geographical area it will still be relevant to, are complex issues. This is highly problematic for career exploration processes where time (now) and place (near here) are generally important. Many social media services facilitate real-time or near-real-time communication. Information is generally served to users in ways that emphasise recent activity. Furthermore, people's use of social media is based on the development of networks of people who share a characteristic or interest. Geographical location remains a powerful characteristic that commonly drives the creation of links in social networks. The value of time- and place-specific web content has been identified by recruiters, who have been engaging job-searchers through using services that make use of new technologies' capacity to recognise time and space.
- *Located in the cloud:* The way in which both applications and data are stored and delivered to the end-user is changing and is increasingly located off-site and with third-party providers. For organisations, cloud computing has big implications for the management of technical infrastructure. The locus of infrastructure moves away from a single organisation towards the internet. For individuals, cloud computing opens up some possibilities that are likely to have implications for career. The ability to create online

spaces that can serve as repositories for data, to move data easily between learning and work environments, and to be able to share these resources with others, has the potential to mainstream the idea of the e-portfolio.

- *Free or almost free:* The cost of publication and development has dropped, enabling a wide variety of resources to be delivered at a much lower cost than in the past. The cost of publication falls dramatically once there is no longer a need to produce a physical object. It also falls when you remove editorial and selection processes and just allow everyone to publish everything. This movement to free or nearly free services has a number of implications. Perhaps most importantly, it creates a culture of "publish and then filter"rather than "filter and then publish". When publication costs money, a lot of energy needs to go into sorting out what is worth publishing. But when publication is free or nearly free, there is no need to do this. The ability to publish freely also provides an opportunity for individuals to market themselves, their ideas and their skills more easily to employers and investors. The profile of the private sector careers consultant/expert/blogger has been increased by the possibility of low-cost publication.
- *Diverse and integrated:* The internet is increasingly integrated into a range of technologies across our lives. In particular, the convergence between telephone and web technologies opens up new opportunities for career learning. The internet is no longer constrained by conventional desktop computing. The growth in internet-focused peripatetic devices has been enormous and looks set to continue (ITU-D, 2010). The ways that people are accessing online services are becoming both more diverse and more integrated. Users can pull content off the web to their telephone or TV, just as they can integrate their Sat Nav or fridge into their computer (O'Hara and Shadbolt, 2008, p.15). Particularly relevant to career learning is the way in which mobile technologies can facilitate situated learning in a variety of authentic contexts. This raises a range of possibilities, such as using mobile learning approaches to support and facilitate learning during work experience for example.
- *Games:* Computer gaming is gaining increasing penetration across society. It is important that the potential of both commercially produced games and bespoke educational games are explored for the purpose of career learning. Gaming has enormous penetration with young people, but has also increasingly moved into mainstream (adult) culture. Purpose-built games have been utilised in education for a number of years. However, there is a growing discussion about the learning that takes place through mainstream computer games

(Royle, 2009). Games-based learning can take place within the context of a formal curriculum, but is more likely to take place outside it. Careers educators have certainly been using face-to-face games and simulations for years to enable people to experiment safely and to explore decisions and transitions (Jamieson, Miller & Watts, 1988).

ROLE OF THE CELL PHONE IN COMMUNICATION NOWADAYS

Two billion people on the planet use cell phones, according to James Katz, professor of communication at Rutgers University. In fact, there are more cell phone subscribers in the United States than there are people, according to an October 2011 study, underwritten by CTIA, the International Association for the Wireless Telecommunications Industry. Cell phones have become so ubiquitous in our culture that Dr. Katz teaches a class on the social aspects of mobile communication. Smartphones, PDAs and other similar devices present the user with a variety of ways to communicate with the touch of a finger.

WRITTEN COMMUNICATION

Where the land line telephone somewhat took the place of written communication when it became commonplace, the cell phone and its texting capability sees Americans communicating, again, with the written word. The difference today, though, is that where once eloquence and verbosity were considered social graces, brevity now reigns. A different "language," composed of acronyms, and now known commonly as text-speak, began online in the early days of email and online chat rooms. An even more abbreviated form spread with the advent of the personal pager, which only allowed seven characters — a phone number — to be entered. Text speak was refined with the widespread use of the cell phone, which initially allowed up to 160 characters.

Electronic Communication

Many Americans use their cellphones to communicate via the internet. Mobile internet capability allows them to check into social media websites, such as Facebook and Twitter, read communications from their friends and colleagues and send quick replies. At times, the communication isn't text, but a photo, completely removing the need for words to convey a response, feelings and even location. Cell phones with broadband capabilities are also frequently used to communicate with others via email.

Verbal Communication

While the sole purpose of the first telephone was to provide a way to speak to someone not in our presence, the cell phone has moved far beyond this concept. In fact, Ian Condry of MIT studies cell phone use in Japan and finds that less than one-third of Japanese cell phone usage involves verbal communication. In some countries, however, it may be the sole way of communicating verbally and maintaining relationships. In certain countries where young men and women are forbidden by religious or cultural norms to communicate in public, the cell phone allows them to speak with one another freely, and without recrimination. A Rutgers University study finds that students speak to their parents more often since the advent of the cellular telephone.

Considerations

Rather than allowing users to "reach out and touch," as one land line telephone company's slogan touted, one study shows that the cell phone insulates Americans more than ever before. For some, communication with those not present takes precedence over interacting with those in the same room. Young people use their cellphones as a method of self-expression, with the color, size and style of the phone often taking precedence over its technical and communication capabilities.

USING ONLINE PROFESSIONAL INFORMATION FOR CAREER DEVELOPMENT

Current, accurate occupational information is a crucial component of successful career decision making, and more and more career information is now available online. To assist career education and career development practitioners in helping students use this information, this *Practitioner File* discusses the role of occupational information in career decision making, addresses some concerns about online information, and takes an in-depth look at some national online information tools provided by America's Workforce Network. Additional website resources for occupational/career information are described.

The Role Of Occupational in order

Career development involves the acquisition of self-knowledge about interests, abilities, and attributes; educational and occupational exploration; and planning that combines self-knowledge and occupational knowledge in making career choices. "Ready access to usable career information is a crucial component of successful career decision making". In the National Career Development Guidelines, skills for understanding and using career information are one of 12 competencies integral to career development (CareerTEC 2000).

The terms "career information," "occupational information," "labor market information," and "work force information" overlap. Broadly defined, these terms encompass the full range of information about labor market conditions and trends, including employment and unemployment, industry and occupational employment and wages, labor market projections, individual workers or job seekers, and providers of education and training (Sommers 2000). Also included is information about specific industries, occupations, and careers, such as qualifications, job duties, educational requirements, working conditions and work settings, compensation, employers, and current job vacancies (O'Reilly 2001; Sommers 2000).

Individual needs for and uses of this information vary: elementary school students need it to develop an understanding of the world of work and the relationship between personal attributes and interests and career choice; through secondary and postsecondary education, students need it to make informed choices about education, career paths, and specific jobs. The need for this information is lifelong, as the workplace changes and adults make career choices recurringly throughout life:

Workers and prospective workers need not only ask the age-old question "What should I do?" they must also ask, "How should I do it?" and "What should I do after that?" and "How can I prepare for my next work as I do my current work?" This is a far more complex set of questions than "What will I be when I grow up?" (*Blueprint for Life* 2001)

In addition, administrators, teachers, and policymakers use occupational information to make decisions about program offerings, curriculum, and resource allocation and about improving linkages with employers (Sommers 2000).

Recent trends underscore the importance of career information for career development (O'Reilly 2001; Sommers 2000). Certificate programs for career practitioner training now include occupational information courses. Labor market information is being written in simpler language for people other than economists to use. Career information for youth is being produced on the Internet and in youth magazines, such as *The Edge*. Federal initiatives by the U.S. Departments of Education and Labor, such as America's Career Resource Network and O*NET, are making information resources increasingly accessible. Research findings highlight the need for accessible, accurate, well-organized information sources and the critical role of practitioners in guiding individuals in the interpretation and use of this information. Studies of elementary school students (Johnson 2000; Toepfer 1999) have shown that many—

- Have only a shallow understanding of how school relates to the real world and limited awareness of the skills and knowledge needed for career success,
- Are heavily influenced by the belief that schoolwork must be specifically career related to be relevant, and
- Have a limited sense of how to develop skills and knowledge and the usefulness of career information in this process.

Adolescents also need assistance in using career information. Kennedy, Christian, and Bell (1999) concluded that adolescents have unrealistic expectations and inconsistencies in their views for their future. A study of British students found that they base their understanding of careers and work on images from personal contact with a job or work, adults, and the media. In a study of Canadian adolescents (Julien 1999), 60% found it difficult to obtain the career information they needed; 40% had to go to too many places; 40% did not know where to go; and 25% did not feel confident asking for information. Many expressed negative attitudes about career choice software, or the availability of school-based print sources and experienced information overload and emotional barriers or time constraints to using career information.

"Research has consistently shown the value placed on family and friends as sources of useful information," according to Patton and McCrindle (2001, p. 32), who noted that noncollege-bound students rely more on social networks than on school-based information services. In particular, parents have an important influence on career choice. However, although parents seem to understand the usefulness of career information sources for their children's career development, many may not know how to take advantage of these resources (Trusty and Watts 1996). Choice of information sources and perception of their usefulness varies depending on such factors as age, gender, socioeconomic status, family circumstances, and cognitive maturity (Jordan and Pope 2001; National Career Development Association [NCDA] 2000; Patton and McCrindle 2001; Trusty and Watts 1996).

Career information should be made available earlier, according to high school seniors in Patton and McCrindle's (2001) study, a need echoed by 69% of adults in the Fourth National Survey of Working America (NCDA 2000), who said that they would try to get more information if starting over. Almost all of the adults in Lucas' (1999) study felt they needed more information not only about occupations, but also on their own interests and skills, a stage of the career development process that should have taken place in elementary school.

However, simply accessing career information is not enough, as other research points to the need for intermediaries such as information professionals and career development practitioners. British adolescents had the perception that school guidance operates as information on demand: once they expressed an interest, they would be given information on that career area (Foskett and Hemsley-Brown 1999). They were prepared to listen only to information about careers they had already chosen and tended to ignore other information they considered irrelevant to that choice. Roney (1999) discovered that adolescents may overaspire in their career expectations because they reason inaccurately about the ease of attaining high-status jobs. Simply giving students labor market information may be insufficient to foster more realistic ideas about careers. A great deal of career information is available online, raising issues of access and user support. Offer (2000) found that 80 percent of college students would rather gain career information from a computer than from a book or a person, 85 percent would rather use a computer program to help them develop a résumé, and 70 percent thought computers could help them find jobs much faster than would traditional job search methods. However, higher percentages recognized the need for personal assistance when those online sources didn't satisfy them. And what of those without online access? NCDA (2000) sees the beginning of a "career information digital divide" that could disadvantage some in the labor market; age, minority group membership, and education level are associated with access to career and job information.

Canada's *Blueprint for Life* (2001) and several U.S. state curricula that use the National Career Development Guidelines (Barbieri 1998a,b; CareerTEC 2000; Ohio Department of Education 2000a,b,c) outline competencies related to career information for elementary, middle school, secondary, and postsecondary/adult levels The competencies increase in complexity as career development continues throughout life. Examples of what individuals should know and be able to do in regard to career information at each level are as follows:

Elementary Education: Understand And Use Career Information

- Explore the work of family and community members
- Identify occupations by people, data, things
- Discover how interests, knowledge, skills, beliefs, and attitudes relate to work roles
- Learn how parents, friends, and relatives can provide career information
- Explore work roles and settings of interest
- Explore the concept of work information and how parents, relatives, adult friends, and neighbors can provide this information

Middle School: Locate, Understand, And Use Career Information

- Identify ways occupations are classified
- Identify occupational groups for exploration
- Use school and community resources to identify occupational groups and work roles
- Discover how skills, knowledge, and attitudes can be transferable from one work role to another
- Identify community employment sources.

Secondary Education: Locate, Evaluate, And Interpret Career Information

- Explore the educational and training requirements of various work roles
- Acquire knowledge of classification systems and career ladders
- Discover how key personnel in selected work roles could become information resources and/or role models Explore how trends and work opportunities in various economic/work sectors affect the nature and structure of work roles

- Understand how a variety of factors (e.g., supply and demand for workers, demographic changes, environmental conditions, geographic location) affect work opportunities
- Understand how labor market information (profiles, statistics, etc.) should be used when making life and work decisions
- Explore a variety of work alternatives

Postsecondary/Adult Education: Locate, Evaluate, And Interpret Career Information

- Identify and use career information resources (computer and Internet-based career information delivery systems, print, media, mentors)
- Identify available work opportunities with respect to one's set of work skills, knowledge, and attitudes
- Understand how to assess the reliability of career information
- Explore opportunities for self-employment
- Assess information and evaluate its impact on one's life/work decisions

In addition to activities that help individuals acquire these competencies, O'Reilly (2001) suggests that career information can be a valuable asset in career decision making only if it is grounded in meaning for the user. She suggests that practitioners guide individuals in "interrogating" the information, using an additional set of questions beyond the in-depth analysis of personal aptitudes, values, and interests and the study of what job or occupation description best matches those personal interests and aptitudes:

- Does the present situation represent the future situation?
- What predictions are reliable?
- Where can I find further information to support or refute these predictions?
- Is a particular article biased to make an argument for a lobby group?
- How will consumer behavior affect opportunities?
- What public policies will impinge on this career?
- How many workers are already in this field?
- How many new workers will be needed?
- How will technology change the industry?
- What are the paths that seem possible from this position?
- Is this occupation mobile across the country?
- From which training programs do employers prefer to hire?

Considerations And Concerns In use *Online Information*

"For many, the Internet has become the vehicle of choice for…accessing various career services" (Kirk 2000, p. 146). A number of factors have contributed to the widespread use of online career information, including (ibid.)—

- The increase in the number of computer-assisted career guidance systems;
- The widespread availability and use of the Internet; and
- A growing number of adults interested in further education and career information.

The array of websites featuring online career information is burgeoning. The sites can be classified into two general categories: career planning and career information. Career planning sites feature such services as self-assessment, occupational information, career exploration, and career decision making. Career information sites, on the other hand, focus on providing information related to employment trends, educational opportunities, employers, and job search (ibid.).

The availability of online career information is paradoxical. Many benefits are associated with greater accessibility to career information, but there are also limitations. Chief among the benefits is greater availability of additional resources to more audiences. Individuals with disabilities, for example, can frequently access online resources more easily than traditional sources. Online resources can also be used to supplement face-to-face contact with career development professionals. Furthermore, they can complement existing print and computer-based assessment resources, and resources that may have been cost-prohibitive to purchase can now be accessed online.

Despite these advantages, however, some limitations and problems are associated with the availability and use of online career information resources. Issues surrounding online career information can be grouped into several categories:

- User proficiency. In order to make effective use of career information available online, users must be proficient in using technology and navigating the Web. Even when the user knows what he or she is looking for, resources on the Internet are organized differently "from those on an orderly free-standing database. The initial experience of the Internet is often expressed as serendipity, or more negatively, trial and error" (Offer 2000, p. 34). Individual users may need guidance in locating appropriate information.
- Quality of information. Quality of information is an area closely related to the quantity of information. The Web is an open system with no quality control. Anyone can mount a site containing career information. Users must become savvy in assessing websites by asking themselves questions about the developer, the content, the layout, and so forth. Another element of quality relates to the availability of computer-assisted career guidance systems through the Web. Is information on the reliability and validity of the assessment available? In many cases, little or no data are provided on an assessment's reliability and validity, making it difficult to

judge whether an instrument is measuring what it is supposed to (Sampson and Lumsden 2000).

- Equity of access. Although the Internet makes career information more widely available, concern exists that some groups may not have access to it. Individuals with limited financial resources, for example, may have difficulty gaining access. Will inequitable access to online career information mean that some groups and individuals will continue to suffer inequality in employment and career opportunities?
- User readiness. Individuals vary greatly in their readiness for career decision making. Those with a high degree of readiness will benefit more from online career information. Individuals with low readiness may experience some difficulties because they need more personalized assistance, and a site may not provide access to help for clarifying directions, asking questions, or interpreting results. Also, individuals using online career assessments may erroneously believe that the instruments will provide a quick and appropriate solution to their career dilemma. If an individual cannot benefit from the type of self-help assistance that is available through online career sites, it would be beneficial for sites to have information about accessing personalized assistance (Sampson and Lumsden 2000).
- Appropriateness of instruments. Some career assessments available online were developed for use in stand-alone computer systems and data resulting from their use as online systems may not be valid or reliable. In addition, some online assessments that are used in the self-help mode were developed for use with a professional career development specialist.

Two sets of standards developed by NCDA address many of the issues raised here and provide helpful guidance to consumers using online career information. The first, "Guidelines for the Preparation and Evaluation of Career and Occupational Information Literature" (NCDA 1991), addresses a number of areas such as accuracy of information, bias and stereotyping, and use of information. These guidelines are useful in assessing both print and online resources. The second, "NDCA Guidelines for the Use of the Internet for Provision of Career Information and Planning Services" (NDCA 1997), is designed for career counselors but is helpful to consumers of online career information. The guidelines can be used to assume a critical stance when accessing online career information.

America's Career Kit and O*NET Online

Two national online occupational and career information tools are available to a variety of users for a variety of purposes. *America's Career Kit* (2001), a career development resource developed and managed by a federal/state partnership led by the U.S. Department of Labor (USDOL), consists of four interrelated websites:

- *America's Career InfoNet* helps users explore different occupations.
- *America's Learning eXchange* helps users investigate education and training opportunities.
- *America's Service Locator* helps users search for local services.
- *America's Job Bank* helps users look for jobs and employers find skilled workers.

*O*NET OnLine* (*What Is O*NET?* 2001) provides interactive Internet access to USDOL's Occupational Information Network (O*NET), a comprehensive database of worker attributes and job characteristics that can be used for career exploration and counseling, education, employment, and training activities.

Both America's Career Kit and O*NET OnLine are part of "America's Workforce Network" (2000), USDOL's nationwide system of work force development organizations providing information and services for individuals managing their careers and employers seeking skilled workers. Another important part of America's Workforce Network is "America's Labor Market Information System" (2000), whose standardized federal, state, and local labor market information and tools are used by both America's Career Kit and O*NET OnLine. This article examines how America's Career Kit and O*NET OnLine can be used for career education.

America's Career Kit

America's Career Kit (Morman 2000) seamlessly integrates job listings, resumes, career information, and learning resources in four websites:

- America's Career InfoNet accesses national, state, and local labor market data, occupational information, and career planning information.
- America's Learning eXchange is a clearinghouse of training and education offerings posted by registered providers.
- America's Job Bank is a nationwide, searchable databank of job openings and resumes.
- America's Service Locator provides maps and directions to service providers.

America's Career InfoNet

America's Career InfoNet (ACINet) is designed to help users make informed career decisions using local, state, and national occupational, demographic, and labor market information and a variety of career information resources ("About Us" n.d.; "Frequently Asked Questions: General Information" n.d.). ACINet shares a set of Career Tools (described later) with America's Learning eXchange.

In addition, four tabs on ACINet's home page allow users to perform quick searches for specific career information:

- General Outlook—Users can generate lists of the top 25 occupations nationwide in 5 different categories (fastest growing, most openings, largest employment, declining employment, and highest paying) by education levels. Users can then select an occupation and a state to generate an Occupation Report with summary state and national wage and trend information and links and a description of the occupation; links in the Occupation Report are available for detailed wages, detailed trends, and industry trends.
- Wages & Trends—Users can also directly generate an Occupation Report with wage and outlook information either by selecting one of 22 job families, or groups of occupations, and selecting from a list of occupations in that job family or by doing a keyword search.
- What It Takes—Users can select an occupation and generate a report of knowledge, skills, and abilities the occupation requires, as well as tasks and activities typically performed (taken from O*NET information); typical education and training for the occupation and links to education institutions in a selected state; and websites with information on the occupation.
- State Information—Users can obtain demographic and economic information by generating a State Profile. From the State Profile, users can generate a General Outlook for the state; links can provide additional information about state employers, resources (career/labor market information, education, cultural recreation), and occupational licensing requirements. Users can also link to America's Service Locator to search for local service providers.

The Career Resource Library links users to a wide range of information and resources both at America's Career InfoNet and on the Internet—job and resume banks, job search aids, occupational information, relocation information, and state resources. Users can either browse the library by categories or search by keywords. Of particular interest for career exploration is the set of nearly 300 downloadable Career Videos, which provide a brief overview of selected occupations.

America's Learning EXchange

America's Learning eXchange (ALX) helps users locate career development, training and education, and employment resources they need to be competitive in the work force (*About ALX*2000). ALX provides a searchable database of profiles on 350,000 offerings—courses, seminars, workshops, degree- or certificate-granting courses, and training materials—available face to face, online, in conferences, via self-study, on TV and video, and by other methods. ALX shares a set of Career Tools (described later) with

ACINet. Users can access Courses, Programs, and Seminars tabs at ACINet's home page to locate offerings. All three tabs use the same form and procedure for searching; users must do a separate search for each type of offering. Users can search either by entering a keyword or by selecting from a list of subjects (e.g., Basic Skills, Conservation and Environment, Legal and Compliance, Therapy and Counseling). Next, users select a state and a delivery method (e.g., classroom, web-based/asynchronous training network, computer-based/ CD-ROM, or all methods). Results are displayed in a list with links to a description of the offering, linked to more information on the provider. Users can also narrow the list by entering additional search criteria (e.g., subtopics, type of provider, date of offering, difficulty level, credit/continuing education unit, and certification.)

Three other tabs allow more specialized searches. The Providers tab allows users to search for offerings by keyword, state, and delivery method. The Accreditation tab allows keyword searches either by name of accrediting agency or by program; it also allows subject area searches (e.g., Allied Health, Engineering and Architecture). The result is a list of agencies that accredit providers and programs. From that list, users can generate a list of all providers and/or programs accredited by the agency. The Certification tab allows searches by keyword or by subject matter. Results are displayed by credentialing institution and certificate name with links to descriptions of certificates; descriptions provide links to ALX courses and programs leading to the certification and to all certifications offered by the credentialing institution.

Shared Career Tools At ACINet And ALX

The Employability Checkup helps users determine if they can find a similar job at similar pay. Users select a state and a local area, educational level, and wage level; a job title or job family that represents their occupation; and an industry. The resulting Employability Profile indicates current employment and expected growth in the selected occupation and industry; local, state, and national unemployment rates; and educational attainment and current wages typical for the occupation. A summary rates occupational growth, industry growth, local unemployment, state unemployment, level of training, and user's current wage. Links are provided to America's Job Bank, America's Service Locator, other state career resources, and general job search information.

The Licensed Occupations tool helps users search for state and federal licensing requirements for a selected occupation. Information is searchable by occupation, state, or agency. Users exploring careers should begin by selecting the occupation they are considering; next, they should select the state in which they plan to practice the occupation, as well as all federal agencies. Search results list the occupation selected, the state and federal licenses required (with links to further descriptive information), and the licensing board (with address and telephone number).

The Employer Locator allows users to identify potential employers (not job openings) from a database of nearly 12 million U.S. employers; reports provide contact and other general information on employers. Users can search by industry, by occupation, or by employer name. Searching either by industry or by occupation involves a gradual process of simultaneously refining three parameters—occupation, industry, and geographic region.

The Financial Aid Advisor helps users identify financial assistance programs for which they might be eligible. Users first answer questions to determine their status as an independent or dependent student; then they provide income, tax, and other information to identify possible program eligibility. Links are provided as appropriate to the online Free Application for Federal Student Aid for grants, loans, and work study; to information on the user's state financial aid programs; and to information on other financial resources such as veteran's assistance, Temporary Assistance for Needy Families, or the Workforce Investment Act. Finally, users are linked to the Scholarship Search Page where they can search by type of award, level of study, vocational goal, and gender, affiliation, and geographic restrictions.

The Career Exploration tool helps users identify their individual needs and use the elements of America's Career Kit, as well as other resources, to meet those needs:

- Explore work and learning options helps identify potentially satisfying occupations and the education and training needed and available. They can learn what career information is; how, when, and why they might use it; and where they can find it, including links to state career information systems (CISs). Users can review the general occupational outlook to identify appropriate career options and review occupational requirements for skills, education and training, and credentialing. Users can examine more detailed state and national outlooks for occupations, as well as current job openings. Links are also provided to other information resources—professional and trade associations, labor unions, government agencies, private organizations, and guidance and counseling associations and services.
- Gain skills to get a job or get a better job helps users identify specific education and training resources they can use to acquire the skills needed for an occupation. Resources include state CISs and information on a variety of options (e.g., federal training programs, colleges and universities, employer-based training, internships, and school-to-work). Finally, users can search for specific classroom-based, work-based, and Internet training offerings and training providers.

- Find financial assistance for education and training identifies state and national sources of financial aid and provides an overview of eligibility requirements and application procedures. A link allows users to access the Financial Aid Advisor (described earlier).
- Find job search resources and services to help you get a job links users with resources and services they can use at different steps in finding a job. Users can follow links to state CISs and state and national occupational information resources on confirming job targets, writing a resume, preparing for an interview, negotiating job offers, relocating, employment law, and finding career counselors and job search professionals.
- Find possible jobs and employers helps users identify positions and employers in the occupation for which they are preparing. Links to America's Job Bank allow users to search for listed job openings and obtain other job search services. Users can also link to ACINet's Career Resource Library to identify potential employers and to review relevant occupational information.

America's Job Bank

America's Job Bank (AJB) is designed to help job seekers find jobs and employers find workers. Job seekers can search a database of 1 million jobs; by registering, they can also create and post an online resume and set up an automated job search. Employers can register to search a database of almost 400,000 resumes, post job openings, and set up an automated resume search.

Job seekers can search for jobs using five tabs on AJB's home page:

- Quick Search allows users to browse among all job listings in one of 22 job families (nationally, internationally, or within a 50-mile radius) or all jobs within a 50-mile radius.
- Job Title allows users to browse a list of the job titles within a job family, select a job title, enter a ZIP code, and select a radius.
- Keyword allows users to search by job title or skill; the Job Title, Certificates/Licenses Required, and Type of Education Required fields of all listings will be searched.
- Military allows those with a military background to enter a Military Occupational Code, rank, and branch and search for equivalent civilian jobs within a selected radius.
- Job Number allows users to recall and review job listings found previously by entering the nine-character Job Number.

Users can refine search results by specifying other location options and other advanced options (new listings only, level of education or experience required, duration, and salary range). Search results include job title, date posted, job location, and listing company, with links to a description for each job title. Job descriptions include links to state profiles and wage and trend information for the type of job. A link on AJB's home page also allows users to search for jobs and employment information in a selected state.

America's Service Locator

America's Service Locator allows users to search for service providers, including One-Stop Centers. Users first enter information on their own location and then choose either Nearest One-Stop or Office Search; Office Search results include not only One-Stop Centers but other centers (e.g., senior centers) or the agencies that sponsor them. Contact details, a map, and driving directions are available for each center listed. A planned Service Search is not yet available.

O*NET OnLine

O*NET OnLine provides access to O*NET occupational information for students; people seeking first jobs, new jobs, or better jobs; and people who are choosing or changing careers. Users can explore these categories of occupational information ("Using O*NET" 2001; "What Is O*NET?" 2001):

- Skills—basic skills (e.g., math, reading) and cross-functional skills (e.g., problem identification, resource management)
- Generalized Work Activities—general job behaviors occurring on many jobs (e.g., organizing, planning, and prioritizing work)
- Interests—broad areas of personal interest (e.g., realistic, investigative)
- Work Styles—personal characteristics and values connected to work (e.g., achievement, initiative, recognition)
- Work Context—physical and social factors that influence the nature of work (e.g., interpersonal relationships, physical work conditions)
- Organizational Context—characteristics of an organization that influence how people do their work (e.g., social processes, personnel resources)
- Experience and Training—levels of preparation needed for occupations, including overall work experience, on-the-job training, and education

Users can access occupational information at O*NET OnLine in different ways:

- Find Occupations allows users to find occupations to explore by searching for a keyword, occupational title, or occupational code or by browsing in one of 23 job families. The resulting brief description for the occupation includes links to an occupational Snapshot and Details.
- Skills Search allows users to search for occupations that require selected skills, including basic skills, complex problem-solving skills, resource management skills, social skills, systems skills, and technical

skills. Users can select an occupation from the resulting list to explore.
- Related Occupations allows users to select and explore other similar occupations.
- Snapshot provides summary information about important aspects of a selected occupation—experience requirements (overall experience, on-the-job training, education), worker requirements (basic and cross-functional skills, general knowledge), worker characteristics (abilities, interests, work values), occupation requirements (generalized work activities), occupation-specific information (tasks), and occupation characteristics (labor market information, occupational outlook, wages).
- Details indicate the level of importance for each of eight variables for a selected occupation—knowledge, skills, abilities, interests, work values, work context, tasks, and work activities.

Private organizations and public agencies can build add-on software and other products that use O*NET as a foundation to meet a range of work-related needs ("What Is O*NET?" 2001). For example, O*NET could be used to cluster occupations based on skills, knowledge, or job tasks; improve counseling or assessment products and processes; align curricula to meet current workplace needs; develop job orders and position descriptions; and improve hiring decisions. Table describes some add-on uses of O*NET and O*NET OnLine.

CURRENT APPLICATIONS OF WEB TECHNOLOGY TO MAINTAIN CAREER DEVELOPMENT

Technological change has the potential to alter radically the ways in which individuals engage in career exploration processes. As discussed in Section 3, new technologies are changing the ways that individuals interact with information, with each other and with the labour market. This section will go on to explore the ways in which organisations are responding to these trends within the context of supporting career development. Much of the practice discussed here is emergent and is included to give a flavour of the kinds of innovations that are currently being explored.

How organisations are supporting career development activities, it is important to recognise that public-sector careers services are not the only providers of relevant web content and services. The environment provided by the web has served to lower the costs for individuals and organisations seeking to publish career-related material. It has also blurred the distinctions between, on the one hand, the trained, public-sector, impartial careers advice, and on the other, services offered by a host of private-sector organisations. It is possible to see this proliferation of career-related information as a threat to the integrity and viability of career IAG. However, it is perhaps more fruitful to see the burgeoning market in online careers information, advice and guidance as an opportunity. In order to make the most of this opportunity, traditional careers services will need to think carefully about their role in this increasingly crowded marketplace. Policy-makers will also have to think about how far the market can be shaped and managed.

Organisations, services and individuals have harnessed the opportunities generated by technological advances to repackage their services, refine and develop their offer, and respond to their markets. Technology has generally been used to help meet client demand in one of three ways:

1. To deliver information;
2. To provide an automated interaction; or
3. To provide a channel for communication.

Where technology is used to deliver information, it serves a range of functions. It can, for instance, recreate the careers library by supplying information about jobs and courses. This can increase clients' access and remove the space limitations that plagued the conventional careers library. However, this kind of technology also provides an opportunity to improve the quality of information, to harness the linked nature of the web to draw in external resources (such as employers' sites) and to provide a more media-rich experience through the use of pictures, audio and video.

Where technology is used to develop an automated interaction, there are a range of opportunities. The use of technology can automate the initial exploration and diagnostic elements of the usual advice and guidance service: for example, it can facilitate psychometric, matching and reflective tools, and perform some initial diagnostic tests. Technology can also be used to support people to develop their career learning skills: for example, through games and simulations that can be used to provide an interactive way of exploring the worlds of learning and work. These technologies therefore both promote user-control and self-reliance, but also can automate some of the more routine aspects of the guidance process, so allowing professionals to focus on offering higher-level support to clients.

Finally, there is a range of tools that facilitate communication and interaction between people, usually but not always at a distance. These technologies include the telephone and email as well as a range of technologies that enable telephone and email communications to be more effectively managed. Such technologies can make professional support to individuals more accessible, and are also being used to build communities of learning.

THE LIBERATION OF CAREER INFORMATION

As with information in many other fields, it is increasingly easy to find a wide range of information to support career development online. Information that might traditionally have required a special trip to the careers adviser or the careers section of the public library is now readily available to anyone with a web connection, and increasingly to anyone with a mobile phone.

This move to provide information freely via the internet reflects wider trends in information provision, but has also been explicitly driven by government in order to support widening access to learning and employment opportunities. For example, the Action Plan for the 14-19 Prospectus and the Common Application Process (DCSF, 2009) has required all areas to develop an electronic prospectus to collate and present complex information about opportunities in learning and training in a format which, they argue, appeals to young people. In some areas, complementary sites have been developed that bring together a wider range of learning opportunities for both young people and adults. In the East Midlands, for example, the Your Future East Midlands website (www.yourfuture-eastmidlands.co.uk) links career destination with current level of learning in order to plot progression pathways and course availability.

Information about different occupations can also be found on a number of sites, including the Prospects graduate careers database (www.prospects.ac.uk), the Careers Advice Service's Advice Resources (www.advice-resources.co.uk/) and the Connexions Direct service (www.connexions-direct.com/jobs4u/). Currently these present largely text-based information, although there are some moves to develop more dynamic and multi-media content.

The sites of public-sector careers services provide authoritative and impartial resources. They often link to the sites of employers and recruitment organisations, which also provide a rich source of occupational information that can support career exploration. For example, the PricewaterhouseCoopers recruitment site (http://www.pwc.com/uk/en/careers/index. jhtml) contains pictures of current employees, a blog written by recent recruits, and up-to-date information about salary, training and vacancies, alongside tips about careers and employability.

Recruitment sites such as Monster (http://www.monster.co.uk/) also contribute to the rich information environment that is available for career exploration. These sites are somewhat more "objective" than employer sites, but because they are commercially based, they are still not fully impartial or comprehensive in the way that a public-sector careers service can aim to be.

The limitations of text-based resources have resulted in the generation of media-rich forms of information dissemination. Thus, for example, iCould (icould.com) offers hundreds of short films describing the career journeys of a cross-section of individuals. Careersbox (www.careersbox.co.uk) offers a similar service, but providing job profiles rather than career journeys. U-Explore (http://www.u-xplore.com/) brings several of these elements together with its development of job profiles, with fact sheets about salary, entry requirements and so forth, alongside video clips, and 360° representations of a typical workspace. Alongside information about course and occupational opportunities, there is information about the labour market. This includes trend information giving a strategic overview developed by the respective Sector Skills Councils, accessible via a portal managed by the University of Warwick (www.guidance-research.org/future-trends).

Individuals engaged in career exploration are also likely to engage with existing vacancy information. This is available in a myriad of places including individual employers (most of whom have a "Vacancies" space on their websites), recruitment agencies and newspapers. In the public sector, job opportunities are available via JobCentre Plus (www.jobcentreplus.gov.uk), while the National Apprenticeship Service has a national Vacancy Matching Service (www.apprenticeships.org.uk/).

The individual has the possibility of accessing a rich range of information for career exploration on the web. Where this information is provided in ways that make it easy to aggregate, to comment upon and to organise through the creation of folkonomies, the pluralism of the web is likely to add substantial benefit.

The use of RSS by the researcher careers organisation Vitae (www.vitae.ac.uk) enables the organisation to provide dynamic content (news, events, blogs, etc) in the form of feeds that can be picked up and utilised by users in whatever context they want (email updates, via a portal such as Netvibes or via an aggregator like Google Reader). The organisation creates content and identifies the content of others, but users are not required to come to the organisation's website to access this content. Similarly, the Manchester University Careers Service uses Twitter as a push technology updating subscribers to changes in the website and providing vacancy information. The use of services such as Twitter allows user to access this information easily through mobile technologies.

Careers information, too, is widely available on the web, and organisations have successfully created a range of ways to disseminate it successfully to users. If they are able to find ways to embrace ideas like community, collectivising knowledge and supporting users who wish to aggregate information, it is possible that this can become increasingly effective. In order to do this, however, organisations are likely to have to give up some control over their content and to need to facilitate individuals to access their information in a variety of new ways. These moves are likely to blur the distinction between what is information (that does not require the intervention of another human) and what is communication.

The main challenge for the user in negotiating this information is the difficulty of organising and evaluating the vast range of information sources. The ability to analyse the bias, subjectivity and partiality of a range of information sources is crucial to successfully navigating online career information. Engagement with technologies and practices such as aggregation, engaging with social networks and the social creation of folkonomies can support individuals in their interrogation of career information.

AUTOMATED COMMUNICATION WITH CAREER RESOURCES

Resources have been developed that take the provision of information further, to offer ways to encourage career learning and career literacy for anyone who chooses to engage with the technology. These are self-managed, self-directed resources that facilitate enhanced self-awareness, and knowledge of occupations and career choices, but which do not require the involvement of a guidance professional.

The ability to take an automated test to increase your self-awareness and make suggestions about your appropriate fit into the labour market is appealing to many individuals. These kinds of tests have been used by both careers services and recruiters in a face-to-face context but are increasingly available via the internet. Some are well-known (such as the Myers Briggs test). Others are less well-known but nevertheless based on academic research: these include the BBC Personality Test (www.bbc.co.uk/science/humanbody/mind/surveys/personality/index.shtml) which is a personality test based on the OCEAN five-factor model for classifying personality traits. A more directly career-based test is offered by the Self-Directed Search (SDS) (www.self-directed-search.com) which provides support for career decision-making based on the theories of John Holland. Interestingly, the SDS is based on a pay-per-test model and marketed towards individuals looking for support in career changing as well as towards young people choosing a career path.

For the non-specialist, it can be difficult to evaluate the validity of the large number of psychometric, matching and guidance tools that are available online. Some are embedded in social networking sites such as Facebook and are traded virally between users: for example, the What Careers Suit Your Personality Type application on Facebook. As with the information sources, the ability to evaluate the provenance and authority of an online tool is likely to be essential. The development of digital literacy and critical information skills will be discussed further in Section 5.

Some companies are building increasingly sophisticated resources that link databases of job types with outcomes from career exploration tests, and relate these to qualification requirements and then to learning providers. Prefinio (www.prefinio.co.uk) developed by Morrisby (who developed the widely-used Fast Tomato) have established what they describe as: "A fully integrated careers education and guidance system that helps individuals of all ages find their ideal career, as well as showing them the routes they could take to achieve their goals. Suggestions are based on an accurate assessment of a person's qualities; their aspirations, expectations, skills and preferences. These attributes are carefully analysed and linked to a comprehensive database of careers, courses, colleges and universities."

Diagnostic tools have been developed to support career learning in other contexts. For example, CXL Ltd have developed C-Xtra EU– an on-line diagnostic tool that allows learners to reflect on their learning needs and prompts them to seek support from their tutors if necessary. Games represent another form of automated interaction, which enable individuals to explore different careers and occupations through simulations. Some of these are developed by career companies to encourage career-related learning in an engaging manner. In formal education, the Real Game (www.realgame.co.uk) has been one of the most successful role-play games devised for an educational setting. The Real Game is now available via the internet and is designed to be played in a classroom setting. Students learn about a range of occupations, the qualifications needed and how occupational choice impacts upon lifestyle, as well as some broader learning related to adult roles and responsibilities. A similar resource is The Drive of Your Life (www.driveofyourlife.org) which encourages career exploration through a driving-based game.

Games companies also develop products that, intentionally or otherwise, facilitate career learning. Board games such as The Game of Life which was published in its original form in 1960 has been repackaged as a CD-Rom version since 1998. More recently, the latest edition of Sims 3: Ambitions provides players with the opportunities to explore a variety of careers (firefighter, architect, investigator, ghost-hunter, stylist or doctor) in a simulated world.

Different in aim and tone to games are computer-assisted simulations. Typically these create a full-automated environment for an individual to explore. An example might be a medical simulation in Second Life, generated primarily for training purposes, but also offering the potential to enable people to explore the working life in a particular occupation.

Some sites have developed very sophisticated means of bringing together simulations or gaming environments with career information and recruitment. An American Army game brings an online gaming environment which includes profiles of roles and YouTube video clips of real soldiers, alongside the army recruitment site goarmy.com which itself has a wide range of multi-media information, chat forums and links to recruiters.

Increasingly, games provide a mix of automated interaction and communication with other people. The Sims example discussed above not merely offers interaction with computer-generated characters, but provides a role-playing environment where individuals can interact with each other. As with the discussion of information giving, the development of new technologies is increasingly blurring the boundaries between automated interactions and communication. Some games like The Sims offer a mix of the two, but others such as Second Life and other Multi-User Domains (MUDs) offer simulated environments populated exclusively by avatars controlled by real people. This has the potential to provide further opportunities for career learning.

The applications described here bring together some of the aspects of new technologies, such as the aggregation of diagnostic results with items drawn from occupational type datasets, and also games that blend high-quality simulated environments with the potential for both automated and virtual interactions to support career learning.

Communication

The new technologies continue to develop new ways to interact with other people. There are a wide range of technologies that provide opportunities to transfer existing career guidance practices online in ways that can overcome barriers of distance or access through the use of email, text chat, video conference or simulated environments. However, it is also important to note that many new technologies are changing the way that people communicate, by extending social networks and facilitating interaction across space and in a time frame that hitherto has been both un-imagined and un-achievable. Career guidance has the opportunity to engage with the complex array of new ways to communicate in order to define new practices and paradigms for the delivery of guidance.

It is possible to classify the range of different communication tools into three main categories: those that facilitate one-to-one, one-to-many and many-to-many forms of communication.

One-to-one

The interaction of professional adviser to client via a face-to-face interview remains an important tenet of practice amongst career guidance practitioners. However, that practice can be effectively translated to communication over the telephone and via the internet as "an opportunity to give more guidance to more people, more often, at a distance" (Offer, 2004). Offer examined a range of channels that could be used to do this, such as web chat, email, on-line discussion fora and message boards. Six years later, the range of communication channels has increased further, but the nature of one-to-one e-guidance remains broadly similar.

One example of how a careers service has innovated with the use of one-to-one communications can be found in the virtual Connexions service in Milton Keynes which encourages access to the service via an appealing virtual website. In the HE sector, personalised advice is being delivered through a range of communication media including for example both audio and visual feedback on interview technique, and using real-time chat facilities to give e-guidance.

Within the field of career guidance there has been some theorisation of online or e-guidance practice. There has also been considerable interest in developing a theory of online practice amongst the mainstream counselling and psychotherapy community. As the career guidance field continues to develop its approaches to online guidance, it may have valuable lessons to learn from practice in the counselling and psychotherapy world.

One-to-one communication does not necessarily have to be with a careers professional to support career learning. Horsesmouth provides a system to support the development of informal mentoring relationships to aid career exploration. It is also possible to use existing professional networking tools like LinkedIn to identify and manage one-to-one communications with recruiters or potential mentors. Finally, it is likely that much peer-to-peer career support is now conducted via a range of technologies.

A particularly specialised form of one-to-one communication is offered through the e-portfolio. E-portfolios are tools through which the outputs of learning and experience can be retained to aid reflection and communication. E-portfolios therefore provide a communication tool between an individual and a range of other parties (e.g. learning tutors, peers, mentors/mentees and potential employers). E-portfolios have been found to support learner reflection (Zubizarreta, 2009), the development of competencies and career learning, but existing e-portfolios have tended to be linked to a particular organisational context. When the learner moves on, access to the e-portfolio has been lost. In principle, access to all-age careers services could enable access to be maintained (Watts, 2010). There is also considerable effort going into exploring inter-operability between various e-portfolio tools (JISC, 2009). However, Cohn & Hibbitts (2004) argue that institutionally based e-portfolios need to be replaced with what they call Lifetime Personal Web Space (LPWS). There is potential for this vision to be realised through a range of web tools. Given the opportunities presented by these new tools for career learning, the question moves from being a technical one to a pedagogic one, with teachers and career practitioners now needing to find ways to encourage the adoption and sustainable use of this wide range of tools.

One-to-many: many-to-one

One-to-many communications are conventionally conceived of as broadcasts. One person (probably the professional) has something to say and the role of the many is simply to listen. However, technologies like blogs and video-sharing sites provide one-to-many communication channels with additional opportunities for the many to respond. This can create a blurring between activities that are primarily about information giving and those that include more interactive advice or guidance.

Blogs, as an example of one-to-many career IAG, are used by both the public and private sectors to provoke thought and raise questions for discussion, as well as for offering more basic forms of information and advice. Private-sector careers professionals tend to use the blog as a means of establishing credibility as part of their marketing strategy. An example here is www.careerealism. com, which blends the blog with video clips offering advice, twitter and an invitation to a career counselling service. Similarly, www.herecomestheboss.com offers careers advice and insight into getting a job and working in a range of sectors but with most of the content as video clips. Public-sector careers services use blogging and Twitter to share information about particular learning or job opportunities, or to share ideas on how to think about career and career learning. Again, higher education professionals have adopted these technologies in advance of others, perhaps because they have been able to assume higher levels of digital access and literacy for longer. Blogs allow their authors to share ideas, and also give them some power to manage and edit the discussions that follow from their initial posts.

As with one-to-one communications, one-to-many communications can take place between a wide range of different career stakeholders. Accenture provides the Accenture Recruiters Blog which as well as broadcasting a range of information to potential recruits also provides an opportunity for job-seekers to engage with recruiters. Similarly, the higher education jobs board jobs.ac.uk provides a range of blogs including the Just Highered blog which traces the fortunes of a recently appointed academic.

In addition to its use by careers professionals, employers and recruiters, blogging is also used as a strategy by job-seekers themselves. A Job Hunter's Blog is subtitled a "blog tracking the progress of one job hunter amongst many" and details the trials and tribulations of going to interviews, receiving rejections and claiming benefits. This kind of blog is clearly designed to raise the blogger's profile, but it also serves as a space for reflection and encourages the evaluation of job-seeking strategies. Finally, it provides a point around which peers can group, engage into discussion and trade strategies.

Many-to-many

Social networking sites facilitate many-to-many communication. It is undoubtedly the case that the peer communities facilitated by social networking sites provide their members with a range of support including peer support. However, this kind of many-to-many communication can be difficult to control and challenging for professionals to intervene in. With most social networking sites, it is relatively easy to create a group to discuss a single theme. Current career-related groups include The Job Seekers Network on Facebook, and the Job and Career Network on LinkedIn. Groups can be spontaneously created by anyone, and frequently fail to reach critical mass.

Many-to-many communication through social networks generally bind together communities of people who share particular characteristics. This is useful for peer support and may lead to connections being made with potential mentors and even employers. However, interventions in such networks by professionals are difficult to manage. The metaphor of the "creepy treehouse" (Stein, 2008) is useful in conceptualising why this is likely to be the case. The creepy treehouse is a space that looks like a fun place to spend time. When kids enter the creepy treehouse, they expect to find other kids like themselves.

But if they discover that it is inhabited by teachers trying to force them to learn, it immediately becomes less appealing or even creepy. This is not to say that intervention by professionals in social networks is impossible: just that it needs to be carefully managed, and access carefully negotiated.

Social networks are not the only form of many-to-many communications. Wikijob provides an alternative conception. It is a resource developed primarily from user-generated content by a community of graduate job-seekers and recruiters. The site is generated through a large number of communications and interactions between its users. It might be possible to view social tagging sites like Delicious as offering a similar form of social communication and collaboration to Wikijob. It would be worth exploring how these kinds of online projects can be used to support career exploration and resource sharing.

EVOLUTION OF MOBILE RADIO COMMUNICATIONS

The rst wireline telephone system was introduced in the year 1877. Mobile com-munication systems as early as 1934 were based on Amplitude Modulation (AM) schemes and only certain public organizations maintained such systems. With the demand for newer and better mobile radio communication systems during the World War II and the development of Frequency Modulation (FM) technique by Edwin Armstrong, the mobile radio communication systems began to witness many new changes. Mobile telephone was introduced in the year 1946. However, during its initial three and a half decades it found very less market penetration owing to high costs and numerous technological drawbacks. But with the development of the cellular concept in the 1960s at the Bell Laboratories, mobile communications began to be a promising eld of expanse which could serve wider populations. Initially, mobile communication was restricted to certain ocial users and the cellular concept was never even dreamt of being made commercially available. Moreover, even the growth in the cellular networks was very slow. However, with the development of newer and better technologies starting from the 1970s and with the mobile users now connected to the Public Switched Telephone Network (PSTN), there has been an astronomical growth in the cellular radio and the personal communication systems. Advanced Mobile Phone System (AMPS) was the rst U.S. cellular telephone system and it was deployed in 1983. Wireless services have since then been experiencing a 50% per year growth rate. The number of cellular telephone users grew from 25000 in 1984 to around 3 billion in the year 2007 and the demand rate is increasing day by day..

CURRENT DAY MOBILE COMMUNICATION

Since the time of wireless telegraphy, radio communication has been used extensively. Our society has been looking for acquiring mobility in communication since then. Initially the mobile communication was limited between one pair of users on single channel pair. The range of mobility was dened by the transmitter power, type of antenna used and the frequency of operation. With the increase in the number of users, accommodating them within the limited available frequency spectrum became a major problem. To resolve this problem, the concept of cellular communication was evolved. The present day cellular communication uses a basic unit called cell. Each cell consists of small hexagonal area with a base station located at the center of the cell which communicates with the user. To accommodate multiple users Time Division multiple Access (TDMA), Code Division Multiple Access (CDMA), Frequency Division Multiple Access (FDMA) and their hybrids are used. Numerous mobile radio standards have been deployed at various places such as AMPS, PACS, GSM, NTT, PHS and IS-95, each utilizing dierent set of frequencies and allocating dierent number of users and channels.

Fundamental Techniques

By denition, mobile radio terminal means any radio terminal that could be moved during its operation. Depending on the radio channel, there can be three dier-ent types of mobile communication. In general, however, a Mobile Station (MS) or subscriber unit communicates to a xed Base Station (BS) which in turn com-municates to the desired user at the other end. The MS consists of transceiver, control circuitry, duplexer and an antenna while the BS consists of transceiver and channel multiplexer along with antennas mounted on the tower. The BS are also linked to a power source for the transmission of the radio signals for communication and are connected to a xed backbone network. Figure shows a basic mobile communication with low power transmitters/receivers at the BS, the MS and also the Mobile Switching Center (MSC). The MSC is sometimes also called Mobile Tele-phone Switching Oce (MTSO). The radio signals emitted by the BS decay as the signals travel away from it. A minimum amount of signal strength is needed in order to be detected by the mobile stations or mobile sets which are the hand-held personal units (portables) or those installed in the vehicles (mobiles). The region over which the signal strength lies above such a threshold value is known as the coverage area of a BS. The xed backbone network is a wired network that links all the base stations and also the landline and other telephone networks through wires.

Radio Transmission Techniques

Based on the type of channels being utilized, mobile radio transmission systems may be classied as the following three categories:

- *Simplex System:* Simplex systems utilize simplex channels i.e., the commu-nication is unidirectional. The rst user can communicate with the second user. However, the second user cannot communicate with the rst user. One example of such a system is a pager.

- *Half Duplex System:* Half duplex radio systems that use half duplex radio channels allow for non-simultaneous bidirectional communication. The rst user can communicate with the second user but the second user can commu-nicate to the rst user only after the rst user has nished his conversation. At a time, the user can only transmit or receive information. A walkie-talkie is an example of a half duplex system which uses 'push to talk' and 'release to listen' type of switches.
- *Full Duplex System:* Full duplex systems allow two way simultaneous com-munications. Both the users can communicate to each other simultaneously. This can be done by providing two simultaneous but separate channels to both the users. This is possible by one of the two following methods.
- – *Frequency Division Duplexing (FDD):* FDD supports two-way radio communication by using two distinct radio channels. One frequency chan-nel is transmitted downstream from the BS to the MS (forward channel).
- A second frequency is used in the upstream direction and supports trans-mission from the MS to the BS (reverse channel). Because of the pairing of frequencies, simultaneous transmission in both directions is possible. To mitigate self-interference between upstream and downstream transmis-sions, a minimum amount of frequency separation must be maintained between the frequency pair, as shown in Fig. 1.4.
- – *Time Division Duplexing (TDD):* TDD uses a single frequency band to transmit signals in both the downstream and upstream directions. TDD operates by toggling transmission directions over a time interval. This toggling takes place very rapidly and is imperceptible to the user.

A full duplex mobile system can further be subdivided into two category: a single MS for a dedicated BS, and many MS for a single BS. Cordless telephone systems are full duplex communication systems that use radio to connect to a portable handset to a single dedicated BS, which is then connected to a dedi-cated telephone line with a specic telephone number on the Public Switched Telephone Network (PSTN). A mobile system, in general, on the other hand, is the example of the second category of a full duplex mobile system where many users connect among themselves via a single BS.

Mobile Call is Actually Made

In order to know how a mobile call is made, we should rst look into the basics of cellular concept and main operational channels involved in making a call. Cellular Concept Cellular telephone systems must accommodate a large number of users over a large geographic area with limited frequency spectrum, i.e., with limited number of chan-nels. If a single transmitter/ receiver is used with only a single base station, then sucient amount of power may not be present at a huge distance from the BS. For a large geographic coverage area, a high powered transmitter therefore has to be used. But a high power radio transmitter causes harm to environment. Mobile communication thus calls for replacing the high power transmitters by low power transmitters by dividing the coverage area into small segments, called cells. Each cell uses a certain number of the available channels and a group of adjacent cells together use all the available channels. Such a group is called a cluster. This cluster can repeat itself and hence the same set of channels can be used again and again. Each cell has a low power transmitter with a coverage area equal to the area of the cell. This technique of substituting a single high powered transmitter by several low powered transmitters to support many users is the backbone of the cellular concept.

Operational Channels

In each cell, there are four types of channels that take active part during a mobile call. These are:

- *Forward Voice Channel (FVC):* This channel is used for the voice transmission from the BS to the MS.
- *Reverse Voice Channel (RVC):* This is used for the voice transmission from the MS to the BS.
- *Forward Control Channel (FCC):* Control channels are generally used for controlling the activity of the call, i.e., they are used for setting up calls and to divert the call to unused voice channels. Hence these are also called setup channels. These channels transmit and receive call initiation and service request messages. The FCC is used for control signaling purpose from the BS to MS.
- *Reverse Control Channel (RCC):* This is used for the call control purpose from the MS to the BS. Control channels are usually monitored by mobiles.

Making a Call

When a mobile is idle, i.e., it is not experiencing the process of a call, then it searches all the FCCs to determine the one with the highest signal strength. The mobile then monitors this particular FCC. However, when the signal strength falls below a particular threshold that is insucient for a call to take place, the mobile again searches all the FCCs for the one with the highest signal strength. For a particular country or continent, the control channels will be the same. So all mobiles in that country or continent will search among the same set of control channels. However, when a mobile moves to a dierent country or continent, then the control channels for that particular location will be dierent and hence the mobile will not work.

Each mobile has a mobile identication number (MIN). When a user wants to make a call, he sends a call request to the MSC on the reverse control channel. He also sends the MIN of the person to whom the call has to be made. The MSC then sends this MIN to all the base stations. The base station transmits this MIN and all the mobiles within the coverage area of that base station receive the MIN and match it with their own. If the MIN matches with a particular MS, that mobile sends an acknowledgment to the BS. The BS then informs the MSC that the mobile is within its coverage area. The MSC then instructs the base station to access specic unused voice channel pair. The base station then sends a message to the mobile to move to the particular channels and it also sends a signal to the mobile for ringing.

In order to maintain the quality of the call, the MSC adjusts the transmitted power of the mobile which is usually expressed in dB or dBm. When a mobile moves from the coverage area of one base station to the coverage area of another base sta-tion i.e., from one cell to another cell, then the signal strength of the initial base station may not be sucient to continue the call in progress. So the call has to be transferred to the other base station. This is called hando. In such cases, in order to maintain the call, the MSC transfers the call to one of the unused voice channels of the new base station or it transfers the control of the current voice channels to the new base station.

WIRELESS COMMUNICATION SYSTEMS

At the initial phase, mobile communication was restricted to certain ocial users and the cellular concept was never even dreamt of being made commercially available. Moreover, even the growth in the cellular networks was very slow.

However, with the development of newer and better technologies starting from the 1970s and with the mobile users now connected to the PSTN, there has been a remarkable growth in the cellular radio. However, the spread of mobile communication was very fast in the 1990s when the government throughout the world provided radio spectrum licenses for Personal Communication Service (PCS) in 1.8 -2 GHz frequency band.

1G: FIRST GENERATION NETWORKS

The rst mobile phone system in the market was AMPS. It was the rst U.S. cellular telephone system, deployed in Chicago in 1983. The main technology of this rst generation mobile system was FDMA/FDD and analog FM.

2G: SECOND GENERATION NETWORKS

Digital modulation formats were introduced in this generation with the main tech-nology as TDMA/FDD and CDMA/FDD. The 2G systems introduced three popular TDMA standards and one popular CDMA standard in the market. These are as follows:

TDMA/FDD Standards

- *Global System for Mobile (GSM):* The GSM standard, introduced by Groupe Special Mobile, was aimed at designing a uniform pan-European mobile system. It was the rst fully digital system utilizing the 900 MHz frequency band. The initial GSM had 200 KHz radio channels, 8 full-rate or 16 half-rate TDMA channels per carrier, encryption of speech, low speed data services and support for SMS for which it gained quick popularity.
- *Interim Standard 136 (IS-136):* It was popularly known as North American Digital Cellular (NADC) system. In this system, there were 3 full-rate TDMA users over each 30 KHz channel. The need of this system was mainly to increase the capacity over the earlier analog (AMPS) system.
- *Pacic Digital Cellular (PDC):* This standard was developed as the counter-part of NADC in Japan. The main advantage of this standard was its low transmis-sion bit rate which led to its better spectrum utilization.

CDMA/FDD Standard

Interim Standard 95 (IS-95): The IS-95 standard, also popularly known as CDMA-One, uses 64 orthogonally coded users and codewords are transmitted simultaneously on each of 1.25 MHz channels. Certain services that have been standardized as a part of IS-95 standard are: short messaging service, slotted paging, over-the-air activation (meaning the mobile can be activated by the service provider without any third party intervention), enhanced mobile station identities etc.

2.5G Mobile Networks

In an eort to retrot the 2G standards for compatibility with increased throughput rates to support modern Internet application, the new data centric standards were developed to be overlaid on 2G standards and this is known as 2.5G standard.

Here, the main upgradation techniques are:

- Supporting higher data rate transmission for web browsing
- Supporting e-mail trac
- Enabling location-based mobile service

2.5G networks also brought into the market some popular application, a few of which are: Wireless Application Protocol (WAP), General Packet Radio Service (GPRS), High Speed Circuit Switched Dada (HSCSD), Enhanced Data rates for GSM Evolution (EDGE) etc.

3G: THIRD GENERATION NETWORKS

3G is the third generation of mobile phone standards and technology, supersed-ing 2.5G. It is based on the International Telecommunication Union (ITU) family of standards under the International Mobile Telecommunications-2000 (IMT-2000). ITU launched IMT-2000 program, which, together with the main industry and stan-dardization bodies worldwide, targets to implement a global frequency band that would support a single, ubiquitous wireless communication standard for all coun-tries,to provide the framework for the denition of the 3G mobile systems.Several radio access technologies have been accepted by ITU as part of the IMT-2000 frame-work.

3G networks enable network operators to oer users a wider range of more ad-vanced services while achieving greater network capacity through improved spectral eciency. Services include wide-area wireless voice telephony, video calls, and broad-band wireless data, all in a mobile environment. Additional features also include HSPA data transmission capabilities able to deliver speeds up to 14.4Mbit/s on the down link and 5.8Mbit/s on the uplink.

3G networks are wide area cellular telephone networks which evolved to incor-porate high-speed internet access and video telephony. IMT-2000 denes a set of technical requirements for the realization of such targets, which can be summarized as follows:

- High data rates: 144 kbps in all environments and 2 Mbps in low-mobility and indoor environments
- Symmetrical and asymmetrical data transmission
- Circuit-switched and packet-switched-based services
- Speech quality comparable to wire-line quality
- Improved spectral eciency
- Several simultaneous services to end users for multimedia services
- Seamless incorporation of second-generation cellular systems
- global roaming
- Open architecture for the rapid introduction of new services and technology.

3G Standards and Access Technologies

As mentioned before, there are several dierent radio access technologies dened within ITU, based on either CDMA or TDMA technology. An organization called 3rd Generation Partnership Project (3GPP) has continued that work by dening a mobile system that fullls the IMT-2000 standard.

This system is called Universal Mobile Telecommunications System (UMTS). After trying to establish a single 3G standard, ITU nally approved a family of ve 3G standards, which are part of the 3G framework known as IMT-2000:

- W-CDMA
- CDMA2000
- TD-SCDMA

Europe, Japan, and Asia have agreed upon a 3G standard called the Universal Mobile Telecommunications System (UMTS), which is WCDMA operating at 2.1 GHz. UMTS and WCDMA are often used as synonyms. In the USA and other parts of America, WCDMA will have to use another part of the radio spectrum.

3G W-CDMA (UMTS)

WCDMA is based on DS-CDMA (direct sequencecode division multiple access) tech-nology in which user-information bits are spread over a wide bandwidth (much larger than the information signal bandwidth) by multiplying the user data with the spreading code. The chip (symbol rate) rate of the spreading sequence is 3.84 Mcps, which, in the WCDMA system deployment is used together with the 5-MHz carrier spacing. The processing gain term refers to the relationship between the signal bandwidth and the information bandwidth. Thus, the name wideband is derived to dierentiate it from the 2G CDMA (IS-95), which has a chip rate of 1.2288 Mcps. In a CDMA system, all users are active at the same time on the same frequency and are separated from each other with the use of user specic spreading codes.

The wide carrier bandwidth of WCDMA allows supporting high user-data rates and also has certain performance benets, such as increased multipath diversity. The actual carrier spacing to be used by the operator may vary on a 200-kHz grid between approximately 4.4 and 5 MHz, depending on spectrum arrangement and the interference situation.

In WCDMA each user is allocated frames of 10 ms duration, during which the user-data rate is kept constant. However, the data rate among the users can change from frame to frame. This fast radio capacity allocation (or the limits for variation in the uplink) is controlled and coordinated by the radio resource management (RRM) functions in the network to achieve optimum throughput for packet data services and to ensure sucient quality of service (QoS) for circuit-switched users. WCDMA supports two basic modes of operation: FDD and TDD. In the FDD mode, separate 5-MHz carrier frequencies with duplex spacing are used for the uplink and downlink, respectively, whereas in TDD only one 5-MHz carrier is time shared between the up-link and the downlink. WCDMA uses coherent detection based on the pilot symbols and/or common pilot. WCDMA allows many performance-enhancement methods to be used, such as transmit diversity or advanced CDMA receiver concepts.Table summaries the main WCDMA parameters.

The support for handovers (HO) between GSM and WCDMA is part of the rst standard version. This means that all multi-mode WCDMA/GSM terminals will support measurements from the one system while camped on the other one. This allows networks using both WCDMA and GSM to balance the load between the networks and base the HO on actual measurements from the terminals for dierent radio conditions in addition to other criteria available.

The world's rst commercial W-CDMA service, FoMA, was launched by NTT DoCoMo in Japan in 2001. FoMA is the short name for Freedom of Mobile Mul-timedia Access, is the brand name for the 3G services being oered by Japanese mobile phone operator NTT DoCoMo. Elsewhere, W-CDMA deployments have been exclusively UMTS based.

UMTS or W-CDMA, assures backward compatibility with the second generation GSM, IS-136 and PDC TDMA technologies, as well as all 2.5G TDMA technologies. The network structure and bit level packaging of GSM data is retained by W-CDMA, with additional capacity and bandwidth provided by a new CDMA air interface.

3G CDMA2000

Code division multiple access 2000 is the natural evolution of IS-95 (cdmaOne). It includes additional functionality that increases its spectral eciency and data rate capability.(code division multiple access) is a mobile digital radio technology where channels are dened with codes (PN sequences). CDMA permits many simultaneous transmitters on the same frequency channel. Since more phones can be served by fewer cell sites, CDMA-based standards have a signicant economic advantage over TDMA-or FDMA-based standards. This standard is being developed by Telecom-munications Industry Association (TIA) of US and is is standardized by 3GPP2.

The main CDMA2000 standards are: CDMA2000 1xRTT,CDMA2000 1xEV and CDMA2000 EV-DV. These are the approved radio interfaces for the ITU's IMT-2000 standard. In the following, a brief discussion about all these standards is given.

CDMA2000 1xRTT: RTT stands for Radio Transmission Technology and the designation "1x", meaning "1 times Radio Transmission Technology", indicates the same RF bandwidth as IS-95.The main features of CDMA2000 1X are as follows:

- Supports an instantaneous data rate upto 307kpbs for a user in packet mode and a typical throughput rates of 144kbps per user,depending on the number of user, the velociy of user and the propagating conditions.
- Supports up to twice as many voice users a the 2G CDMA standard
- Provides the subscriber unit with upto two times the standby time for longer lasting battery life.

CDMA2000 EV: This is an evolutionary advancement of CDMA with the following characteristics:

- Provides CDMA carriers with the option of installing radio channels with data only (CDMA2000 EV-DO) and with data and voice (CDMA2000 EV-DV).
- The cdma2000 1xEV-DO supports greater than 2.4Mbps of instantaneous high-speed packet throughput per user on a CDMA

channel, although the user data rates are much lower and highly dependent on other factors.

- CDMA2000 EV-DV can oer data rates upto 144kbps with about twice as many voice channels as IS-95B.

CDMA2000 3x is (also known as EV-DO Rev B) is a multi-carrier evolution.

- It has higher rates per carrier (up to 4.9 Mbit/s on the downlink per carrier). Typical deployments are expected to include 3 carriers for a peak rate of 14.7 Mbit/s.Higher rates are possible by bundling multiple channels together. It enhances the user experience and enables new services such as high denition video streaming.
- Uses statistical multiplexing across channels to further reduce latency, en-hancing the experience for latency-sensitive services such as gaming, video telephony, remote console sessions and web browsing.
- It provides increased talk-time and standby time.
- The interference from the adjacent sectors is reduced by hybrid frequency re-use and improves the rates that can be oered, especially to users at the edge of the cell.
- It has ecient support for services that have asymmetric download and upload requirements (i.e. dierent data rates required in each direction) such as le transfers, web browsing, and broadband multimedia content delivery.

3G TD-SCDMA

Time Division-Synchronous Code Division Multiple Access, or TD-SCDMA, is a 3G mobile telecommunications standard, being pursued in the People's Republic of China by the Chinese Academy of Telecommunications Technology (CATT). This proposal was adopted by ITU as one of the 3G options in late 1999. TD-SCDMA is based on spread spectrum technology.

TD-SCDMA uses TDD, in contrast to the FDD scheme used by W-CDMA. By dynamically adjusting the number of timeslots used for downlink and uplink, the system can more easily accommodate asymmetric trac with dierent data rate requirements on downlink and uplink than FDD schemes. Since it does not require paired spectrum for downlink and uplink, spectrum allocation exibility is also increased. Also, using the same carrier frequency for uplink and downlink means that the channel condition is the same on both directions, and the base station can deduce the downlink channel information from uplink channel estimates, which is helpful to the application of beamforming techniques.

TD-SCDMA also uses TDMA in addition to the CDMA used in WCDMA. This reduces the number of users in each timeslot, which reduces the implementation complexity of multiuser detection and beamforming schemes, but the non-continuous transmission also reduces coverage (because of the higher peak power needed), mo-bility (because of lower power control frequency) and complicates radio resource management algorithms.

The "S" in TD-SCDMA stands for "synchronous", which means that uplink sig-nals are synchronized at the base station receiver, achieved by continuous timing adjustments. This reduces the interference between users of the same timeslot using dierent codes by improving the orthogonality between the codes, therefore increas-ing system capacity, at the cost of some hardware complexity in achieving uplink synchronization.

Index

N

P

R

S

T